The germ cell history of Rana cantabrigensis Baird

I. Germ cell origin and gonad formation

By

Tso-Hsin Cheng

With 12 figures in the text

Sonderabdruck aus
Zeitschrift für Zellforschung und mikroskopische Anatomie
Fortsetzung des Schultze-Waldeyer-Hertwigschen Archiv für Mikroskopische Anatomie und der Zeitschrift für Zellen- und Gewebelehre
(Abt. B der Zeitschrift für wissenschaftliche Biologie)
16. Band, 3. und 4. (Schluß-) Heft
Abgeschlossen am 3. Dezember 1932

Springer-Verlag Berlin Heidelberg GmbH
1932

ISBN 978-3-662-39294-2   ISBN 978-3-662-40328-0 (eBook)
DOI 10.1007/978-3-662-40328-0

Die Zeitschrift für
Zellforschung und mikroskopische Anatomie

steht Originalarbeiten aus dem Gesamtgebiet der beschreibenden und experimentellen Zellen- und Gewebelehre sowie der Mikroskopischen Anatomie der Menschen und der Tiere offen.

Die Zeitschrift erscheint zur Ermöglichung raschester Veröffentlichung zwanglos in einzeln berechneten Heften; mit etwa 50 Bogen wird ein Band abgeschlossen.

Der für diese Zeitschrift berechnete Preis des Heftes gilt nur zur Zeit des Erscheinens.

Das Honorar beträgt M. 40.— für den 16seitigen Druckbogen.

Die Mitarbeiter erhalten von ihren Arbeiten, wenn sie nicht mehr als 24 Druckseiten Umfang haben, 100 Sonderabdrücke, von größeren Arbeiten 60 Sonderabdrücke unentgeltlich. Doch bittet die Verlagsbuchhandlung, nur die zur tatsächlichen Verwendung benötigten Exemplare zu bestellen. Über die Freiexemplarzahl hinaus bestellte Exemplare werden berechnet. Die Mitarbeiter werden jedoch in ihrem eigenen Interesse ersucht, die Kosten vorher vom Verlage zu erfragen.

Die Herren Autoren werden ferner gebeten, den Text ihrer Arbeiten so kurz zu fassen wie es irgend möglich ist, sich in den Abbildungen auf das wirklich Notwendige zu beschränken und nach Möglichkeit Federzeichnungen (für Strichätzung) zu verwenden.

Alle Manuskripte und Anfragen sind zu richten an
Professor Dr. R. Goldschmidt, Berlin-Dahlem, Kaiser-Wilhelm-Institut für Biologie
oder an
Professor Dr. W. von Möllendorff, Freiburg i. Br., Anatomisches Institut, Albertstr. 17.

Die Herausgeber

Goldschmidt   von Möllendorff

Verlagsbuchhandlung Julius Springer in Berlin W 9, Linkstr. 23/24
Fernsprecher: Sammel-Nrn. Kurfürst 6050 u. 6326. Drahtanschrift: Springerbuch-Berlin
Reichsbank-Giro-Konto und Deutsche Bank, Berlin, Dep.-Kasse C

---

16. Band.   Inhaltsverzeichnis.   3. und 4. (Schluß-) Heft

Seite

Cheng, Tso-Hsin, The germ cell history of Rana cantabrigensis Baird. I. Germ cell origin and gonad formation. With 12 text figures 497

Cheng, Tso-Hsin, The germ cell history of Rana cantabrigensis Baird. II. Sex differentiation and development. With 32 figures in the text 542

Vihvelin, H., Über die Muskelspindeln der Amphibien (Frosch und Kröte). Mit 9 Textabbildungen 597

von Korff, K., Zur Histologie und Histogenese der verschiedenen Zementarten, insbesondere die Beteiligung derselben am Aufbau der kompliziert zusammengesetzten Zähne. Mit 25 Textabbildungen 608

Slonimski, P. und Z. Lapinski, Zur Methodik des histochemischen Nachweises von Hämoglobin und dessen Verbindungen. Mit 4 Textabbildungen . 653

Muggia, Giulio e Lorenzo Masuelli, La citologia della cellula epatica in rapporto alla qualità dell' alimento introdotto. Con 11 figure nel testo 659

Wermel, E. M. und Z. P. Ignatjewa, Studien über Zellengröße und Zellenwachstum. I. Mitteilung: Über die Größenvariabilität der Zellkerne verschiedener Gewebearten. Mit 17 Textabbildungen 674

Wermel, E. M. und Z. P. Ignatjewa, Studien über Zellengröße und Zellenwachstum. II. Mitteilung: Über die Veränderungen der Zellgrößen bei Gewebeexplantation. Mit 8 Textabbildungen 689

Fortsetzung des Inhaltsverzeichnisses auf der III. Umschlagseite.

THE GERM CELL HISTORY OF RANA CANTABRIGENSIS BAIRD.

I. GERM CELL ORIGIN AND GONAD FORMATION[1].

By

Tso-Hsin Cheng,

Zoölogy Department, Fukien Christian University, Foochow, China.

With 12 figures in the text.

(Eingegangen am 21. April 1932.)

Introduction.

It is a familiar fact that the germ cells, which, during early ontogenetic development, appear to be relatively slightly differentiated, become in later stages very highly specialized, after a series of characteristic changes, which prepare them for taking part in fertilization and the subsequent formation of new individuals. The complexities of the structures that appear, and of the processes that occur, in the germ cells immediately preceding and during syngamy, have become more and more obvious with the advance of knowledge. The embryonic origin of the germ cells, together with their early history of development, has been, for a number of years, a subject for speculation and scientific research. The study of sex differentiation is intimately involved in the analysis of the biological nature and basis of sexuality and in the various current problems of sex research. The seasonal variations and activities of the various constituents of the gonads present problems of considerable interest and importance not merely on account of the process of annual gametogenesis, but also because of their bearing on the question of germ-cell continuity.

With the above comparatively important phases of the germ-cell cycle in view, it has seemed desirable to make a thorough investigation of the entire germ cell history, in order to obtain a comprehensive outlook and an adequate understanding of all the various phases involved. The present paper represents a study of this kind on the northern woodfrog, *Rana cantabrigensis* Baird.

While considerable literature has accumulated on isolated periods or phases of the germ cell history of frogs, there is no complete account of the whole history in any one species. From time to time, different investigators, working on different phases of the germ cell cycle in

---

[1] Contribution from the Zoological Laboratory of the University of Michigan.

closely related forms, or even in the same species, have arrived at generalizations, which are directly opposed to one another. Such conflicting views have made apparent the need for a comprehensive and consecutive account of the germ cells throughout the entire period of ontogeny.

The author acknowledges with pleasure his indebtedness to Professor PETER OKKELBERG for advice and encouragement as well as for a valuable training which has made possible whatever contributions this paper may offer. Sincere thanks are also due to Mr. WESLEY CLANTON for much aid in securing material for study, and to other friends and associates for various services rendered during the progress of the work. A part of this investigation was carried on under the tenure of a Graduate Fellowship of the University of Michigan.

Materials and Methods.

Early in the breeding season, frogs were collected from ponds in the vicinity of Ann Arbor, Michigan, and brought to the laboratory. Mating pairs were isolated, and each pair kept in a separate aquarium which was made to imitate, as far as possible, the natural conditions best suited for spawning. As soon as the eggs were laid, the parents were killed and their gonads sectioned and examined carefully for any germinal abnormalities. Tadpoles were reared in medium-sized bacteria dishes. All tadpoles of the same brood were kept separate from other broods. They were fed exclusively on liver, lettuce and hard-boiled egg yolk. A few pieces of *Elodea* were put in the cultural dishes, which served to aerate the water. The room temperature was kept around 20 degrees Centigrade, with occasional variations of about 5 degrees above and below.

The eggs and the tadpoles hatched from them were timed in their development from the moment of spawning. The material was fixed at desired intervals. In this way, a complete series of developmental stages up to and including metamorphosis was obtained. The tadpoles of any given age, especially after hatching, showed considerable variations in the rate of body growth, which variations became more obvious in the course of development. The size or length of the body or of the entire tadpole, was, therefore, not found to be a reliable criterion of age. In our experience, there was no single factor which proved to serve as a satisfactory index for the degree of development attained.

For a study of postlarval development of the germ cells, young, immature frogs were collected afield and fixed as collected. Adult frogs were captured at various times of the year and their gonads removed and fixed immediately.

As to fixatives, BOUIN's and ZENKER's fluids were extensively used, both of which gave very satisfactory results. CHAMPY's, CARNOY's, FLEMMING's strong, and other solutions were employed with fair success.

In sectioning larval gonads, the mesonephroi were generally left attached to them. A few tadpoles of this species were collected in the ponds and their gonads sectioned to furnish comparisons. In no case, was there any detectable difference in the general course of germ cell development between the laboratory and field specimens. In the case of adult specimens, only the gonads were dissected out for imbedding. Most of the adult ovaries were sectioned only in part. Young embryos were difficult to cut, owing to the abundance and extreme brittleness of the yolk material they contained. The difficulty was much reduced when the materials were fixed in solutions containing a large percentage of formalin, and were sectioned within a short time after fixation. Long preservation tended to harden the tissue and alter the normal staining capacity.

All sections were mounted in series and were either transverse or sagittal, rarely frontal. They were cut at 6, 8 or more often 10 micra in thickness.

The staining was invariably HEIDENHAIN's iron-alum hematoxylin. In certain cases where a greater contrast was desired, light-green was used for counterstaining.

Natural History.

Rana cantabrigensis BAIRD, commonly known as the northern wood-frog, is found in abundance in the vicinity of Ann Arbor, Michigan. The frog is small in size, with an average body length of about 40 mm. and is best recognized by a triangular dark or black patch in the ear region immediately behind the eye. This species bears a remarkable resemblance to *Rana sylvatica* LE CONTE, or the eastern wood-frog, and is mainly distinguished from the latter by leg measurement. In *Rana cantabrigensis*, the hind limbs are relatively short, their length to the heel equalling the combined lengths of head and body. In *Rana sylvatica*, the length of the leg to the heel is considerably greater than the total measurement of head and body. Distinctive characters of the two species have been studied and compared by HOWE (1899) and BOULENGER (1920).

In range of distribution, *Rana sylvatica* is confined to the northeastern United States; it is not found west of the Great Plains, south of South Carolina, or north of Quebec, Canada (DICKERSON, 1906). The distribution of *Rana cantabrigensis* is wholly northern, extending from Michigan, Wisconsin, Illinois and Minnesota westward to British Columbia, and northward to Alaska and James Bay (DICKERSON, 1906; RUTHVEN, THOMPSON and GAIGE, 1928).

The northern wood-frogs emerge from hibernation early in the spring. The time of their first appearance is undoubtedly conditioned by the temperature of the air, although not wholly dependent upon it. In

1928, mating pairs were seen on the first day of April. In 1929, egg-masses were first recorded on March 26th. During this season, the males have distinct and deeply pigmented nuptial pads and are very active. They have a sharp, high-pitched voice and their combined croaking is very deafening. Such an intensive chorus naturally tends to bring the frogs together in large numbers. Very often, within an area of 10 square feet or even less, one can find 10 or more mating pairs and many unmated individuals. When suddenly approached, they disappear almost instantaneously in the debris on the bottom of the pond.

It is a noticeable fact that in the spring the male frogs always appear first. At the beginning of the breeding season, one always finds an overwhelming predominancy of males. Out of one collection of 150 early in the season only 2 were females. Among frogs collected at other times of the year, the sexes were approximately equal in number. For an adequate determination of the sex ratio, it would be necessary to examine a large number of specimens from any given frog population, which procedure, however, is beyond the scope of the present investigation.

The eggs of our wood-frogs are laid in masses, which are invariably submerged and nearly always attached to sticks, grass-stems or the like. Eggs kept in the laboratory at a room temperature of 15 to 20 degrees Centigrade hatch in 5 days. In the ponds, the time of hatching is variable, owing to temperature fluctuation. Development is slowed down greatly in cold weather.

Wood-frog tadpoles are omnivorous. They can be fed with vegetables, liver, crumbled bread, hard-boiled egg yolk and other foods. They devour with eagerness any dead or dying fellow larvae. A full-grown tadpole before absorption of the tail in metamorphosis, has a total length ranging from about 40 to slightly over 50 mm. Tadpoles reared in the laboratory metamorphose between 60 and 70 days after the time of egg-laying.

As soon as the breeding period is over, the frogs leave the ponds for the surrounding woodlands. They begin active feeding, which continues throughout the warm season. They are silent most of the time and become more and more solitary in their habits with the oncoming of cold weather. During summer and autumn, they are usually found in moist places, among grasses and mosses and sometimes in wooded localities quite remote from the water. They hibernate under stones, among dry leaves, beneath mossy logs, and in kindred places. During the months of December, January and February, which are the coldest months in this part of the country, only a few frogs have been collected afield, despite a most strenuous search for them.

Germ Cell Origin and Gonad Formation. Observations.

The germ cells of *Rana cantabrigensis* are first recognizable in tadpoles at the time of hatching. Prior to this stage, there are no satisfactory means yet available for identifying the germ cells, if they do exist as such. The descriptive data to be recorded below are chiefly concerned with that portion of the body, which is more or less directly involved in the early development of germ cells and gonads. The developmental history will be presented in stages for convenience of description. The processes described as occurring in any given stage or in successive stages may overlap in more or less degree, but the general scheme of development is precisely the same.

1. Embryos 4—5 days old and 5—7 mm. total length.

The notochord is a conspicuous structure composed of greatly vacuolated cells. Lying directly beneath it, is the small hyponotochord, the compact structure of which contrasts with the vacuolated notochordal tissue. The WOLFFian ducts are well marked, with small but definite lumina. In the region of the lateral mesoderm, there appears no apparent differentiation into somatic and splanchnic layers. The cardinal veins and the dorsal aorta are readily recognized as irregular and thin-walled vessels, containing a few spherical, yolk-laden cells, which later lose their yolk contents, and acquire the shape and character of red blood corpuscles. The paired cardinals lie medially to, and in contact with the WOLFFian ducts of the respective sides. The archenteron appears as a small cavity, longer dorso-ventrally than laterally. The entodermal tissue is densely packed with large and mostly oval-shaped yolk bodies, staining intensely black with iron-hematoxylin, and this color clings to them even after extraction from other parts of the body. Diffused throughout these large masses of yolk, there are small granules which stain less deeply, and appear brownish in color. Owing to the abundance and crowded condition of the vitellary materials, other cellular structures become much obscured, and cell boundaries indistinct and difficult to determine. In the mesoderm, on the contrary, yolk is much reduced in amount, the yolk spherules are loosely arranged, and measure, in most cases, less than 5 micra in diameter, while those in the entoderm are, on the average, 2—3 micra larger.

Underneath the dorsal aorta, the roof of the archenteron is elevated above the level of the vitelline entoderm as a distinct ridge (fig. 1), which extends nearly the entire length of the yolk-mass. As regards their morphology, the cells of this entodermal ridge closely resemble one another and are, in all appearance, identical with the primitive entodermal cells. They are readily distinguished from the adjoining mesoderm not only by the position they occupy but also by the amount of yolk they contain and the character of the yolk granules.

The entodermal ridge subsequently plays an important rôle in the development of germ cells. It is from a certain definite portion of this ridge that the germ cells later appear to be derived. This implies, and is indicative of, the possibility that the prospective germ cells, previously

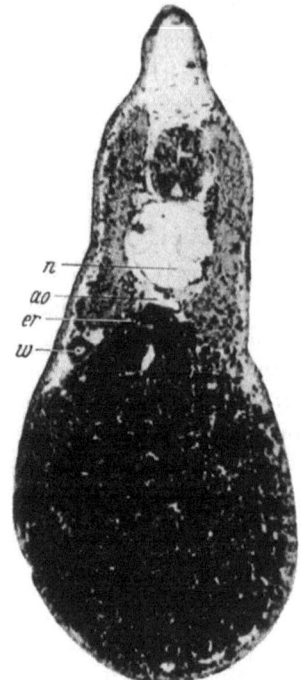

Fig. 1. Transection of an embryo 5 days old and 4 mm. total length, showing the entodermal ridge immediately underneath the dorsal aorta. *ao* dorsal aorta, *er* entodermal ridge, *n* notochord, *w* Wolffian duct.

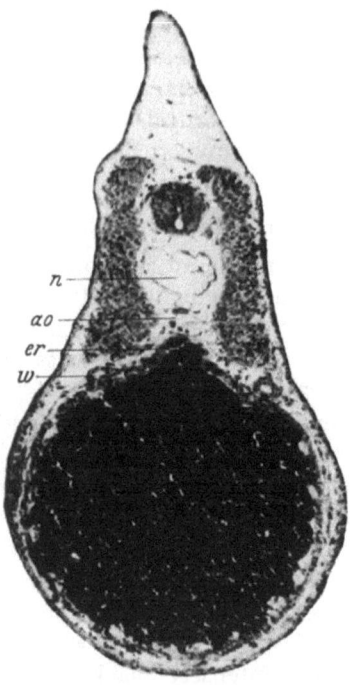

Fig. 2. Transection of a larva 6 days old and 7 mm. total length. The germ cells are in the process of separating from the yolk entoderm, forming the germ cord. *ao* dorsal aorta, *er* entodermal ridge, *n* notochord, *w* Wolffian duct.

to their appearance as distinct structures, might be present among the cells of the entodermal ridge, either potentially or actually as such, though morphologically unrecognizable.

2. Larvae 6 days old and 7—8 mm. long.

In the middle trunk region, the entodermal ridge becomes the seat of certain remarkable developments. The lateral plates of the mesoderm, one from each side of the body, have grown dorsomedially, extending above the yolk-mass, and approximating each other to form the anlage of the dorsal mesentery. With the formation of the coelomic cavity and its extension toward the median axis of the body, the mesenterial anlage becomes a well-defined structure. The germ cells, hitherto located in the entodermal ridge, are being separated from the latter to become included in the mesenterial anlage, forming a structure, which may be designated as the median germ cord (fig. 2). The cord is first formed anteriorly, and the process progresses backward as in the usual order of embryonic differentiation. The formation of this structure and the processes concerned in its formation are among the most significant features in the early history of germ cells.

It may be emphasized that our evidence is demonstrative of the fact that the germ cells are actually entodermal in their early location. The lateral plates, before their union medially to form the future dorsal mesentery, never have contained any germ cells nor cells comparable to them as regards the character and amount of the yolk content. Moreover, by tracing a number of serial sections, the germ cells are frequently observed in the actual process of separating from the vitelline entoderm to become embedded in the apposing splanchnic layers of lateral mesoderm.

Once in the mesoderm, the germ cells stand out with great prominence, owing to the sharp contrast they bear to their surrounding elements. They are very similar to the underlying yolk cells, from which they can be distinguished mainly or perhaps only, by location.

The germ cells occur more or less *en masse*, without any regular arrangement or organization. They are not uniformly distributed in the germ cord, being generally more abundant in its middle portion than elsewhere. They show no segmentation of any kind, although occasionally there are one or more sections, which do not show germ cells in the usual position. This, however, can not be regarded as of a metameric nature, as it is a purely local condition and incidental to germ-cell distribution.

The cytological features of the germ cells are very indistinct on account of their yolk contents. Their only reliable characteristic during this period is their general position in relation to other structures and the abundance of their yolk platelets, the grouping of which may be taken, in some instances, as indicative of the size of the cells.

The caudal end of the germ cord is about 300 micra anterior to the level of the cloacal openings of the WOLFFian ducts. The length of this cord varies in different individuals, depending apparently, in part, on the degree of development which the respective individual has attained at the time of killing. Several measurements indicate an average length of 450 micra, with variations of 100 micra more or less. Few scattered germ cells have been encountered cephalad or caudad of the germ cord, included among the mesodermal cells of the splanchnopleure or of the developing mesentery.

Simultaneously with or immediately preceding the formation of the germ cord, the cardinal veins, originally confined to the region medial to the WOLFFian ducts, have extended laterally above and beyond the latter. Thus, each cardinal becomes recognizable as two portions, fairly distinct, though still continuous with each other. The portion laterad to the WOLFFian duct is the incipient renal portal vein, while the more medial portions form the venae revehentes, which later are united forming the inter-renal portion of the postcava.

3. Larvae 7 days old and 8—9 mm. long.

The rapid enlargement of the coelom brings about a proportionate elongation of the dorsal mesentery. As the latter elongates, it carries the germ cord farther and farther away from the place of its first appearance. This process is invariably more advanced anteriorly than posteriorly, and the extent of its progress may be correlated with the degree of development which the mesentery has attained.

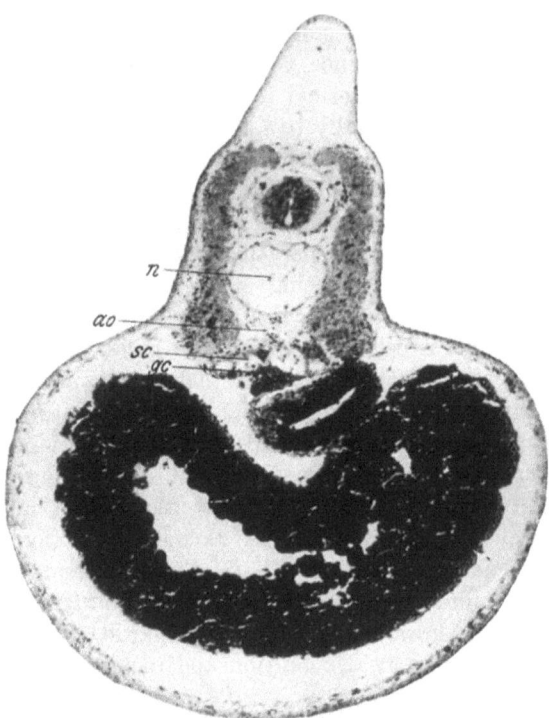

Fig. 3. Transection of a larva 7 days old and 8 mm. total length. Note the distinct germ cord at the base of the dorsal mesentery. *ao* dorsal aorta, *gc* germ cord, *n* notochord, *sc* subcardinal vein.

Throughout the process of shifting, the germ cells show no indication in their morphological constitution that they are capable of independent movement. They are confined to the loose mesenchyme of the developing mesentery, and are accompanied by, or rather associated with, small mesodermal cells, which contain relatively deeply-stained nuclei but no yolk. In rare instances, few germ cells, singly or in small groups, are delayed in migration by faulty correlations in the mechanics of growth and development, and are left behind in the splanchnic mesoderm or in the ventral part of the mesentery.

The germ cord eventually comes to lie in an enlarged portion at the base of the gut mesentery, closely adjacent to the aorta dorsally and to the revehent veins laterally (fig. 3). It tends to assume a bilobed appearance, which condition is indicative of the future formation of the paired germ ridges. The germ cells, though far removed from the entodermal cells, still bear a close resemblance to the latter in their cytomorphic features. They appear as large rounded masses of yolk platelets and pigment granules, among which, one can occasionally see portions of the nuclei, which stain lightly, each containing a distinct spherical nucleolus. The number of cells is difficult to ascertain but judging from the differing amounts of the yolk platelets, one is convinced that there is a great deal of individual variation with respect to the germ-cell number.

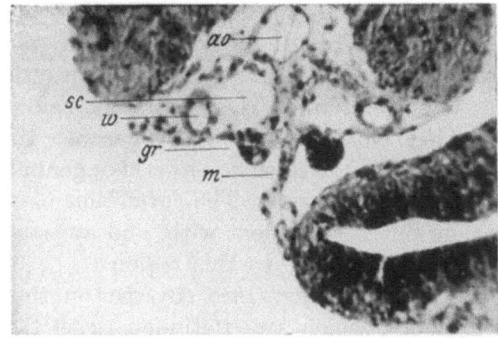

Fig. 4. Transection of a larva 9 mm. total length. The germ ridges are formed, one on each side of the gut mesentery. The subcardinals (revehent veins) lie above the germ ridges, and are approaching each other to form the posterior portion of the postcava. *ao* dorsal aorta, *gr* germ ridge, *m* dorsal mesentery, *sc* subcardinal vein, *w* Wolffian duct.

4. Larvae 8—9 days old and 9—11 mm. long. With the approach of the venae revehentes medially to form the future postcaval vein, the germ cord appears to have undergone a process of segregation. Groups of

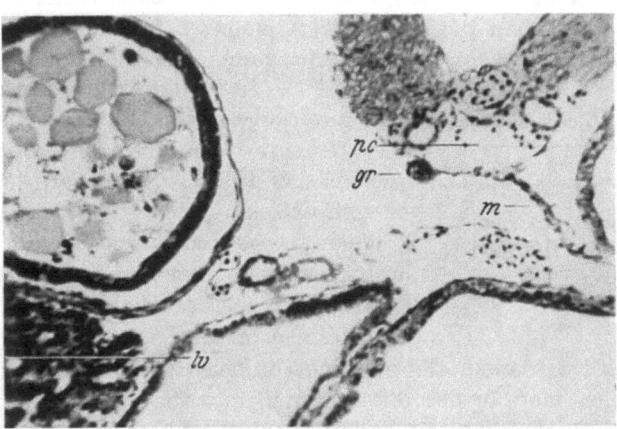

Fig. 5. Transection of a larva 10 mm. total length. It shows the anterior extremity of the right germ ridge, which, in this specimen, extends further anteriorly than the left one. *gr* germ ridge, *lv* liver, *m* dorsal mesentery, *pc* post cava.

germ cells thus segregated, together with their associated mesenchymal elements, move laterally and become aggregated in bulky masses, which project into the coelomic cavity on each side of the mesentery. The processes are gradual and generally proceed from in front backwards.

In many specimens, the germ ridges fully developed in the cephalic portion of the germinal territory are found to converge posteriorly into a median germ-cell mass, which is the remaining portion of the germ cord (fig. 4).

With the progress of development, the projecting structures become eventually established at their full length, and are arranged bilaterally as low longitudinal ridges directly below the venae revehentes, which lie very close to each other beneath the dorsal aorta. These paired ridges may be designated as germ or genital ridges which are the primordia of the germ glands. The formation of these structures under normal conditions is coincident with, and appears to be initiated by, the migration of germ cells to this region.

The germ ridges are covered on their free surfaces by peritoneal envelopes, which are continuous with the general coelomic epithelium, and which are derived from the parietal mesoblast, though appearing, in some instances, to include some of the mesothelium of the mesentery. The ridges are devoid of any regular metamerism, but show local enlargements or thickenings due to the concentration of contained cells. These thickenings, however, do not bear any correspondance to the body somites, nor do they show any constant resemblance to one another in size and shape.

As regards their relative lengths, the germ ridges may be similar or may differ in varying degrees. In the latter case, either one of them may be longer than the other. Fig. 5 shows a further extension of the right germ ridge beyond the anterior limit of the left one. In general, however, the left ridge is longer and extends more anteriorly than the right one. A few measurements taken from the representative specimens of this age group are listed below for comparison.

Lengths of Genital Ridges

1. Left	2. Right
0.43 mm	0.47 mm
0.52 mm	0.36 mm
0.41 mm	0.39 mm
0.46 mm	0.50 mm

There is no indication of any factor or mechanism which may be responsible for the length differences of the germinal structures at this stage of development.

During this period, the vitelline entoderm has lost most of its yolk and becomes more or less differentiated into the alimentary tract. The germ cells, on the contrary, are still packed with yolk substance, though many of them have distinctly shown a rapid diminution of it. In rare instances, one can see the germ-cell nuclei, which are small and rounded, measuring about 9 micra in diameter. They stain very faintly with iron hematoxylin and each contains a distinct basophilic nucleolus. The germ cells as yet show no visible evidence of an approaching mitosis. This condition may be related to their retention of much yolk, a feature rendered more evident and characteristic by contrast with the somatic tissues which have used nearly all their yolk materials in division and differentiation.

Examination of the sections of the developing germ ridges without any knowledge of the previous history of the germ cells might lead one to assume that the germ cells are derived *in situ* from a somatic origin. Such an assumption is untenable, as our evidence clearly demonstrates an early extraregional origin of the germ cells. The germ cells can be traced back without question to their first appearance in the median germ cord. Moreover, they can always be distinctly identified and clearly distinguished from the mesodermal cells among which they lie and through which they migrate. There is no indication in any of the stages so far studied of a possible transformation of peritoneal cells, or the so-called "*Paragonien*" (KUSCHAKEWITSCH), or any other elements of the mesodermal tissue, into the germ cells.

5. Larvae 10—11 days old and 10—13 mm. long.

The revehent veins have been anastomosing in the genital region, forming the inter-renal portion of the postcava, which lies at the basal attachment of the dorsal mesentery, a position formerly occupied by the germ cord. The germ ridges, previously situated below the venae revehentes, have now become attached to the ventro-lateral side of the postcava. The ridges during further growth protrude farther into the body cavity, carrying with them a thin layer of peritoneum as an enveloping fold. Their basal attachment becomes constricted to form the mesogonium. The ridges are, thus, converted into primordial germ glands, which are commonly called indifferent gonads before the time of morphological sex differentiation.

At this stage, the process of yolk absorption is in its full course. The yolk granules which still persist vary considerably in size, the largest ones measuring 7—9 micra. In position, they may be aggregated along the peripheral portion of the cell, surrounding the centrally situated nucleus; or they may be confined to one side of the cell body, with the nucleus on the opposite side. Among the yolk spherules are interposed minute pigment particles, which become more and more distinct as the yolk granules gradually disappear.

The germ cells range from 16 to 24 micra or more in size, and are typically spherical or ovoid in shape, with fairly definite and regular outlines. The nuclear membrane is frequently indistinct because of the vitelline granules present. The nuclei appear more or less rounded, averaging 9—10 micra in diameter, and generally contain one, rarely two or more, basophilic nucleoli of about 2.5 micra in size. The germ-cell protoplasm, in sharp contrast to the contained yolk and pigment materials, stains but feebly and appears pale and homogeneous. Considering all the above features, the germ cells can in no way be confused with the small, compact and relatively darkly stained mesoblastic elements. No cells of doubtful origin nor cells of a possible transitional nature have been observed.

The distribution of germ cells in the germ glands, the relative lengths of and the actual positions occupied by the gonads are indicated in fig. 6. The gonads have steadily increased in length, extending anteriorly to the level of the dorsal attachment of the ligamentum suspensorium hepatis and even further beyond. A peculiar arrangement of structures in this region prevents the forward growth of the right gonad, a phenomenon which becomes more apparent with the progress of development, and is further described and discussed in a subsequent section of this paper.

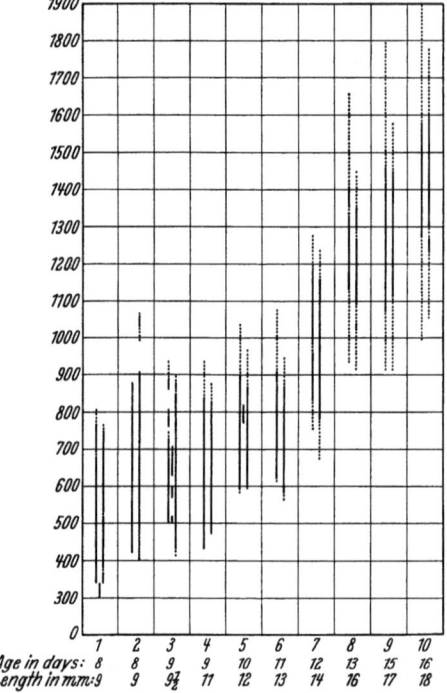

Fig. 6. Diagrammatic representation of germinal structures (germ cord and ridges in earlier stages; germ glands in later stages) found in a series of tadpoles ranging from 8 to 16 days in age. The basal line indicates the level of the cloacal openings of the Wolffian ducts. The ordinates are distances in micra, measured anteriorly from that level. In each column, a pair of genital ridges or of germ glands are represented, left and right corresponding to the left and right sides of the reader. (For further explanation see text.)

The growth of the gonads at this stage is evidently due to a rapid increase in the amount of their somatic tissues. While stroma cells and cells of the gonad peritoneum show many nuclear similarities, there is no direct indication that they are genetically related. In our materials, the evidence is convincing that the stroma cells are derived extragonadally from the nephrogenic tissue. In serial sections, one can readily follow the migration of cells from the mesonephric blastema toward and actually into the gonads. This migration, though occurring on both sides, is not a symmetrical process. It appears segmental in nature, but is often very irregular in this respect.

6. Larvae 10—20 days old and 12—22 mm. long.

The primordial gonad during the earlier part of this period is a small and compact structure. In a typical transverse section, it appears as a solid cord, containing a few germ cells usually near its distal periphery, and numerous stroma cells elsewhere. The whole organ is enclosed by a delicate peritoneal investment, which is often too thin to be readily recognizable.

The stroma of the gonad increases steadily by mitoses of the pre-existing mesenchymal elements, and by continued migration of similar cells from the nephrogenous tissue. The stroma cells are dispersed throughout the organ with no definite organization. They constitute the anlagen of the rete cords, or at any rate, form a greater part of the materials necessary for the development of such. In structure, these mesenchymal stroma elements are smaller and more compact than the germ cells. They are devoid of definite cellular membranes, appearing, thus, to form a dense syncytium. Their nuclei are distinct, measuring 5 or 6 micra in diameter and containing, besides numerous scattered dark strands of chromatin, one or more nucleoli. These nuclei resemble closely those of the mesonephric cells. The former, however, are smaller than the latter, and, instead of being rounded, are often somewhat flattened owing to mutual pressure in a crowded situation.

The germ cells or gonocytes, as they are often called, show varying degrees of yolk absorption. With the disappearance of yolk, the nucleus, hitherto much obscured, becomes more conspicuous, and gradually increases in size, an indication of nuclear growth, which leads to cell division. There are generally one or two nucleoli, rarely more than two. Attenuated and irregular plumous filaments of karyoplasm with fine and faintly stained chromatin granules scattered on them, are seen radiating from the nucleoli, and apparently surrounding them in the form of an intricate network. The nucleus may be centrally or eccentrically placed, and may be spherical, oval or polymorphic in form. In extreme cases of nuclear polymorphism, different lobules of the same nucleus may appear in one section as two separate nuclei, each with a nucleolus. Such a condition can easily be mistaken for an amitotic division. The cytoplasm of the gonocyte, after the complete loss of yolk platelets, still contains many pigment granules and is otherwise clear and homogeneous.

In the mass of materials studied, we have found no extensive degeneration of gonocytes, nor any regeneration of them from a somatic source. Germ-cell degeneration in our form is a rare and unusual phenomenon.

The gonocytes are each surrounded by a follicle, the cells composing which are usually stretched to an extreme thinness. The follicle nuclei always conform to the general shape of the cell body. They are much flattened, and somewhat spindle-shaped, containing one distinct and few less conspicuous and smaller nucleoli. In these respects, and in their basophilic reaction, they are similar to the nuclei of peritoneal cells. The gonocytes are frequently surrounded on their external surface by delicate peritoneal cells, which, in these cases, have thus assumed the rôle of follicle cells. In some instances, the peritoneal cells appear to have migrated inward from the gonad peritoneum to form a part of the

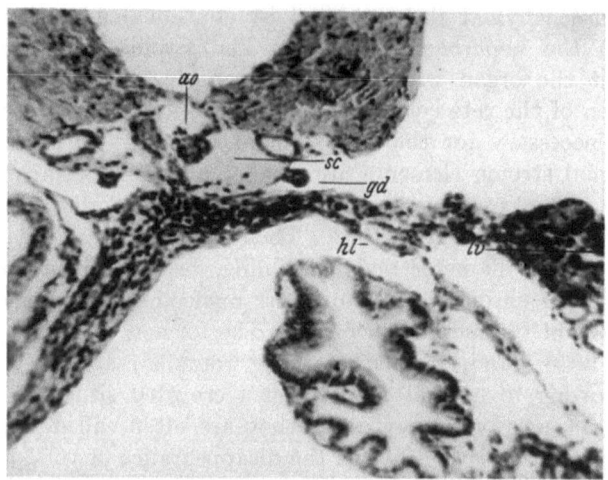

Fig. 7. Transection of a larva 14 days old and 14 mm. total length. It cuts through the anterior extremity of both gonads, showing that neither gonad has any relationship to the hepatic ligament (ligamentum suspensorium hepatis). The material was sectioned from the posterior end forward, so that the right side of the photograph corresponds to the same side of the reader. *ao* dorsal aorta, *gd* gonad, *hl* hepatic ligament, *lv* liver, *sc* subcardinal vein.

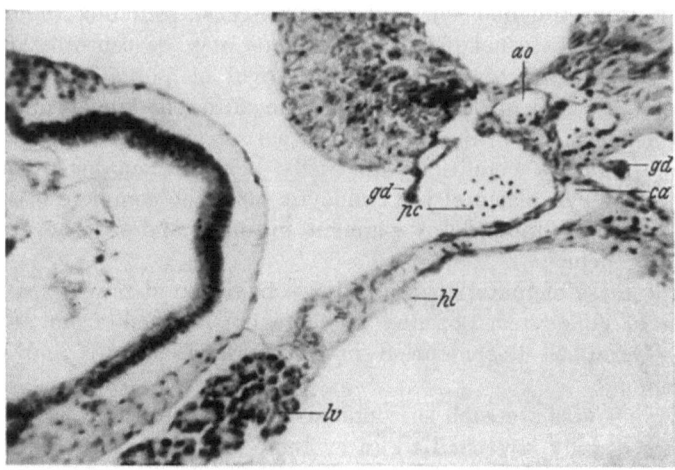

Fig. 8. Transection of a larva 15 days old and 14½ mm. total length. Note both gonads and their relationship to the hepatic ligament. *ao* dorsal aorta, *ca* coeliaco-mesenteric artery, *gd* gonad, *hl* hepatic ligament, *lv* liver, *pc* postcava.

definitive germ-cell follicle. Observations favor a peritoneal origin of the follicle, although the possibility of a stromal derivation is not precluded.

With the advance of age, the germ glands become shifted more and more anteriorly to the cloacal openings of the WOLFFian ducts. This, as shown in figure 6, is not accompanied by any corresponding decrease in the length of the gonads, and is apparently due to the rapid elongation of the body in this region.

By referring to figure 6 again, it may be noted that in most cases, the left gonad is longer than the right one. The difference in length is variable, and is due to further extension of the left gonad either anteriorly or posteriorly or in both directions beyond the limits of the right gland. WITSCHI (1929a) states that in *Rana sylvatica* the asymmetry in the length of gonads is caused by the attachment of the

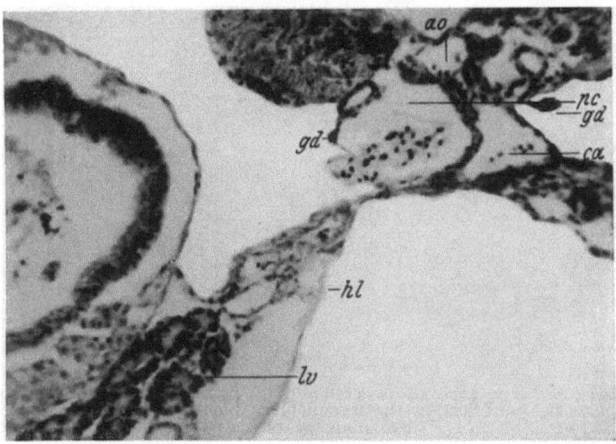

Fig. 9. This transection is 20 micra anterior to that shown in fig. 8. The postcava has extended toward the right side of the body. Note the reduction in size of the right gonad. *ao* dorsal aorta, *ca* coeliaco-mesenteric artery, *gd* gonad, *hl* hepatic ligament, *lv* liver, *pc* postcava.

ligamentum suspensorium hepatis to the left side of the base of the gut mesentery. This, it should be pointed out, accounts for only the difference in the anterior limits of the gonads. In only a few of our preparations, are we able to confirm WITSCHI's results. In many other cases, however, the right gonad has no direct relationship to the suspensorium hepatis (fig. 7 and 8), and is actually obliterated by the lateral expansion of the right revehent vein toward the right side to meet the hepatic portion of the postcava, as shown in figure 9. The blood vessel which lies in the hepatic ligament is a branch of the coeliaco-mesenteric artery which has been mitaken for a "portal" vein by WITSCHI (1929a, p. 240).

The regional concentration of gonocytes is a noticeable feature during this period of gonad development. The unbroken lines in figure 6 indicate the extent of the germ-cell territory; the dotted lines represent the sterile portions. The latter portions are generally confined to the extremities of the gonads. They contain only mesenchymal elements, enclosed externally by a surface peritoneum. The terminal tips of the gonads appear as irregular masses of peritoneal cells.

Throughout the period of formation and development of the primordial gonads, occasional germ cells have been found located outside the germinal structures, within the mesothelial folds of the developing mesentery, or near its dorsal attachment, or embedded in the mesenchyme immediately above the level of the gonad. In earlier stages, scattered gonocytes are observed in the partition wall between the two adjacent

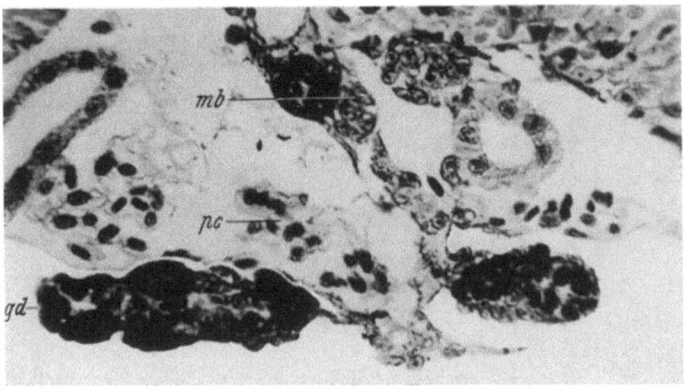

Fig. 10. Transection of a larva 16 days old and 15¹/₂ mm. total length, showing an ectopic germ cell above the postcava in the mesonephric blastema. *gd* gonad, *mb* mesonephric blastema, *pc* postcava.

revehent veins, which have subsequently fused to form the postcava. In rare cases, single germ cells come to lie in the mesonephric tissue (fig. 10). These ectopic germ cells appear normal in structure, with no apparent degenerative changes. They contrast sharply with the surrounding cells which are small and devoid of yolk. Morphological evidence definitely excludes the possibility of their derivation *in situ* from somatic elements.

7. The completed indifferent gonads (fig. 11).

With continued growth and differentiation, the indifferent gonads soon attain their maximum development. The stroma cells, previously irregularly scattered throughout the whole gonad, now show a tendency toward consolidation, as a result of which they become aggregated into compact masses segmentally arranged along the longitudinal axis of the gonad. These solid masses of cells receive additional elements of similar nature, migrating from the mesonephros, and form what may be designated as rete cords. These cellular cords are confined to the medullary portion of the gonad, and the cells composing them are derived, as previously described, from outside the gonad and begin migrating into it as the latter is just being formed.

The number of rete cords ranges from 3 to 7, and varies in different individuals and gonads. It seems to be conditioned by the length of the gonad and the degree of gonadic development. The size of rete cords is proportionate to the width and thickness of the gland in the regions concerned.

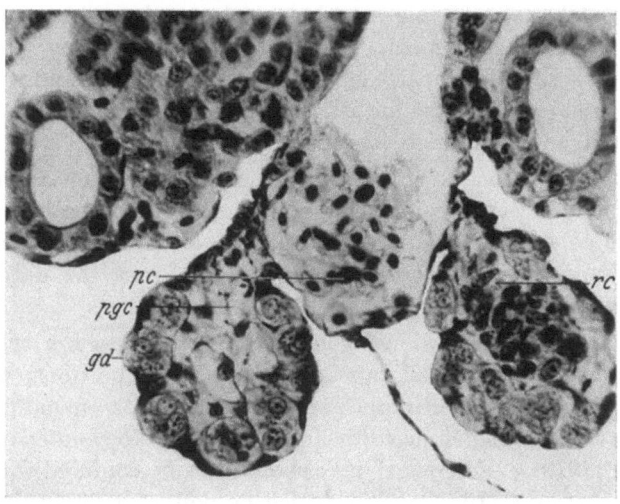

Fig. 11. Transection through indifferent gonads. Note their attachment to the lateral side of the postcava. Both gonads show distinct germ cells in their cortical region. The primary genital cavity is shown in the right gonad. The medulla of the left gonad is occupied by small stroma cells, which constitute a rete cord. *gd* gonad, *pc* postcava, *pgc* primary genital cavity, *rc* rete cord.

The metameric arrangement of rete cords produces a sort of gonadic segmentation, which, however, shows little or no correspondance or relationship to the muscular metamerism. The space intervening between these cords is the so-called primary genital cavity, the lumen of which is mostly occupied, and very often entirely obliterated, by scattered mesenchymal cells, loosely connected by thin protoplasmic filaments. Small vascular vessels with few blood cells ramify through the loose stroma. This genital cavity frequently extends to the rete-cord region, surrounding the distal ends of the cords.

Enclosing the central core of rete cords and inter-cordal "cavities", is an irregular layer of cells, which forms the cortex of the gonad. This cortical layer is most prominent in the regions where germ cells are found in abundance, and is practically absent from the sterile portions. It is generally thicker along the distal periphery of the gland than elsewhere. The germ cells which compose this layer are arranged in a single row, and are isolated from one another by their follicles and scanty stroma materials. At all stages, they remain distinct, and are clearly distinguishable, from the adjoining cells, the so-called *"Paragonien"* (Kuschakewitsch) or *"petites cellules germinatives"* (Bouin). The gonocytes show varying amounts of pigment granules and some of them still contain many or only a few yolk spherules. The germ-cell nuclei contain several distinct nucleoli, and are generally polymorphic, which condition is one of their most characteristic features. The germ cells as yet show no visible morphological sex characters, and may be regarded as indifferent gonia. They soon begin an intense mitotic activity, which results in a rapid increase in their number and a noticeable decrease in size. The first mitotic figures are encountered in a tadpole 16 days old and 19 mm. total length.

The whole gonad is covered by a delicate peritoneal layer. The germ cells lie closely against this covering, but not within it. Proximally, the gonad peritoneum is continuous with the mesothelium of the mesogonium, which supports the organ and forms the only pathway to and out of the gonad.

As a result of the regional concentration of the germ cells as elsewhere described, the gonad can be divided into 3 portions, which may be called progonad, gonad proper and postgonad, corresponding respectively with the *progonalen, gonalen* and *epigonalen Abschnitten* of Kuschakewitsch (1910). "Progonad" here should not be confused with the progonad of Swingle (1921—1926), which he defines as a sexually neutral structure later replaced by the definitive testis, or, in other words, a pre-gonad, or more specifically, a pre-testis. Etymologically "postgonad" is to be preferred to the term *"epigonalen"* of Kuschakewitsch, or "epigonium" of Witschi.

The developmental history of germ cells and gonads thus far described and the terminology adopted for the various structures formed in their development may be reviewed in a diagrammatic scheme as follows:

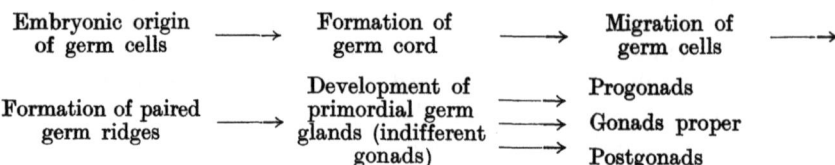

The pro- and post-gonads are practically sterile, containing occasionally single, or a few sporadic germ cells. In later stages, the progonad gives rise to the corpus adiposum, while the postgonad persists as a rudimentary ligament. The gonad proper forms the definitive germ gland, which continues to grow and normally differentiates into a testis or an ovary, or abnormally into an intersexual gonad.

Literature and Discussion.
Origin of Germ Cells.

The early germ cells of vertebrates have been variously described or designated as primitive ova (BALFOUR), primary sexcells (ALLEN), primordial or primitive germ cells (BEARD) and primary gonocytes in English terminology, as Ureier (WALDEYER), Urgeschlechtszellen and Urkeimzellen in German terminology, and as ovules mâles (POPOFF), gonocytes primordiaux, ovules primordiaux, oeufs primordiaux, grandes cellules germinatives, grandes cellules sexuelles primordiales (see BOUIN, 1901; and DUSTIN, 1907) in French terminology.

Concerning the origin of the primordial germ cells in vertebrates, a survey of literature reveals a great diversity of opinion. It is beyond the scope of our discussion to present all the controversy in details. At best, we may briefly review the better known theories on this subject, with regard to the fundamentals involved.

A. Germinal epithelium theory. This theory first advanced by WALDEYER (1870) claims that the primary germ cells or Ureier arise by direct transformation of mesodermal cells in the surface peritoneum of the gonad, or the so-called germinal epithelium. Thus, the germ cells are regarded as differentiated products of the epithelial tissue of the soma. This idea has received the indorsement of most earlier workers and especially of many investigators on germ cells of mammals. Many adherents of this theory have failed to recognize the germ cells previously to their localization in the gonad. Some, however, have observed the primordial germ cells in earlier stages, but maintain that they do not produce sex elements. These cells have been described or referred to as hypertrophied local elements of unknown significance, as abnormal or accidental products of cleavage doomed for ultimate degeneration, or as ordinary somatic cells in preparation for division, or in certain phases of metamorphosis or accentuated metabolic activity, and later participating in the formation of mesenchyme or other somatic tissues (HARGITT, MINOT, VON BERENBERG-GOSSLER, WINIWARTER and SAINMONT, and others).

Among the supporters of WALDEYER's hypothesis, many hold that the definitive sex cells can be and are proliferated by the germinal epithelium throughout the lifetime of the individual until after the cessation of sexual activity (ROBINSON, 1918; ARAI, 1920; E. ALLEN, 1921—1923; SUN, 1923; HARGITT, 1923—1930; PAPANICOLAOU, 1924; BUTCHER, 1927; SWEZY and EVANS, 1929). According to KINGSBURY (1913, 1914), KINGERY (1917), COWPERTHWAITE (1925), and others, the potentiality and capacity of the germinal epithelium for germ-cell proliferation is lost as the gonad attains maturity. Still others, such as B. M. ALLEN (1904) in mammals and BUTCHER (1928, 1929) in lampreys, believe that the germ cells are derived from peritoneal cells only during embryonic stages, or during a limited period of embryonic life, after which time, they increase in number by mitoses and not by continued transformation of somatic cells.

It has also been claimed that the germ cells are produced by a process of differentiation or transformation from the periblastic elements (REINHARD, 1924), blood cells (VALAORITIS, 1879), stroma cells (FOLEY, 1927, among others), epithelial cells (MCGREGOR, 1899; HARGITT, 1923, 1924; et al.), peritoneal derivatives, or other elements of the mesoderm. These various views are similar to the germinal epithelium theory in ascribing to the germ cells a mesoblastic or somatic origin.

B. Early segregation theory. NUSSBAUM (1880), BEARD (1900—1902), and others have advocated the idea that the germ cells are direct products of cleavage, set aside at the very beginning of ontogenesis, and reserved exclusively for this destiny. The germ cells are specific entities, and although being contained in, and protected and nourished by, the body, they remain at all times independent of the latter genetically. This view is in accord with, and gains support from, WEISMANN's work (1885) on germplasm continuity, a hypothesis first proposed by OWEN (1849). Thorough studies of the history of germ cells in many invertebrates have disclosed much evidence in support of the early segregation theory. Keimbahn-determinants have been discovered and described by various investigators, for example, *Chromidien* by BUCHNER (1910) in Sagitta, *Ectosomen* by AMMA (1911) in Copepoda, and pole-disc by HEGNER (1908, 1909) in Chrysomelidae. For a more extensive bibliographic literature, one may consult the work of the last named author (1914). By means of the Keimbahn determinants, it has been claimed possible to identify the germ cells during their early formation and to trace them from their place of origin to the site of gonad fundament. In vertebrates, evidences of an early segregation of germ cells have been found by WHEELER (1899), and OKKELBERG (1921) in cyclostomes, by NUSSBAUM (1880), EIGENMANN (1891—1896), BEARD (1900—1902), WOODS (1902), FEDOROW (1907), DODDS (1910), ALLEN (1909, 1911a), SINK (1912), BACHMANN (1914), RICHARDS and THOMPSON (1921), HANN (1927), and STROMSTEN (1929) in fishes, by NUSSBAUM (1880), ALLEN (1907), KING (1908), WITSCHI (1914, 1929a), SWINGLE (1921, 1926), BECCARI (1921 to 1924), BOUNOURE (1924—1929), and HUMPHREY (1925) in amphibians, by ALLEN (1906), JARVIS (1908), GASPARRO (1908), and JORDAN (1917) in reptiles, by HOFFMANN (1893), NUSSBAUM (1901), RUBASCHKIN (1908), TSCHASCHIN (1910), SWIFT (1914—1916), WOODGER (1925), RICHARDS, HULPIEU, and GOLDSMITH (1926—1928) in birds, and by RUBASCHKIN (1908—1912), FUSS (1911, 1912), VANEMANN (1917), and KOHNO (1925) in mammals.

C. Gonotome theory. According to RÜCKERT (1888), the gonads were derived from a definite portion of the segmental mesoblast of the embryo.

He found most of the "primordial ova" in the somatic mesoderm of *Selachia*, which fact, he believed, gave support to the theory that the reproductive organs of the vertebrates were originally segmented as those in Amphioxus. VAN WYHE (1889) corroborated RÜCKERT, and applied the term "gonotome" to the mesodermal anlagen producing the germ cells.

The theories presented above evidently contradict each other, and are, in many respects, mutually exclusive. Some authors have proposed certain modifications and combinations of the various views. It has even been claimed (WALDEYER, 1901—1903; FELIX and BÜHLER, 1906; et al.) that the theory of early segregation may hold true in vertebrates below the mammals, while in the latter, germ cells are derived from the germinal epithelium or peritoneal derivatives. There are cases on record, in which the germ cells of different sexes in the same species are described as having different modes of origin (WALDEYER, 1870; KUSCHAKEWITSCH, 1910; KIRHAM, 1916). On account of the large accumulation of recorded facts and discordant interpretations, an extensive review of literature would require a disproportionate amount of space and cannot be included here. Important works on the germ-cell origin in amphibians, supposedly based on original observation of the embryonic development of germ cells, may be presented in the following table. The references are listed in chronological sequence; those of the same year are arranged alphabetically according to the authors.

Among the interesting features of this table, it may be noted that there exists in the published accounts a wide divergence of conclusions even on the same species, e. g., *Rana temporaria*. It becomes quite obvious that the apparent differences in results obtained are not always due to actual differences of the material studied, but rather to a noticeable disagreement among the investigators in their explanation and interpretation of the known facts.

As seen in the tabular review of literature, the extraregional origin of germ cells of anurans finds its first mention in GOETTE's classical work on *Bombinator* (1875, p. 831), but the author did not formulate any definite conception of a precocious segregation of germ cells. In *Salamandra maculata*, FÜRBRINGER (1878) evidently recognized the primordial germ cells before they appeared in the germ glands, as clearly indicated in his figure of a 14 mm. embryo (See Tab. II, fig. 18 of his paper). During the time of these writings, the germinal epithelium theory was widely accepted, without, however, any confirmatory evidence from careful observations. Not until 1880, did NUSSBAUM advance a rival theory in which he maintained, largely on theoretical speculations, a blastodermic origin of germ cells and their morphological continuity in successive generations. Many of the subsequent researches have confirmed NUSSBAUM's view in its original or in a more or less modified form.

Table I. *Review of literature on the embryonic origin of the germ cells in amphibia.*
Urodela.

Author	Year	Species	Results
KOLESSNIKOW, N.	1878	Triton cristatus Triton taeniatus	Germ cells are derived from the germinal epithelium
HOFFMANN, C. K.	1886	Triton cristatus	Germ cells are derived from the germinal epithelium
HALL, R. W.	1904	Ambystoma punctatum	Germ cells are derived from the dorso-median portion of the lateral mesoderm (gonotome)
DUSTIN, A. P.	1907	Triton alpestris	Germ cells are derived from the median portion of the lateral plate (gonotome), and from peritoneal cells
ALLEN, B. M.	1911b	Ambystoma Necturus	Germ cells arise from the portion of mesoderm between the future myotome and the future lateral plate
SCHAPITZ, R.	1912	Ambystoma mexicanum	Germ cells are derived from the medial portion of the lateral plate (gonotome)
SPEHL, G. et POLUS, J.	1912	Ambystoma tigrinum	Germ cells are derived from the medial portion of the lateral plate (gonotome) and from peritoneal cells
ABRAMOWICZ, H.	1913b	Triton taeniatus	Germ cells are derived from the entoderm and from peritoneal cells
CHAMPY, C.	1913	Triton palmatus	Germ cells are derived from the mesoderm medial to the nephrotome (gonotome)
BECCARI, N.	1922	Salamandrina perspicillata	Germ cells are derived from the median portion of the lateral plate (gonotome)
BOUNOURE, L.	1924b 1925	Triton alpestris	Germ cells are first found in the entoderm
HUMPHREY, R. R.	1925	Hemidactylium scutatum Triturus viridescens Triturus totosus Eurycea bislineatus Desmognathus fuscus	Germ cells are first found in the medial portion of the lateral mesoderm
BURNS, R. K.	1925	Ambystoma punctatum	Germ cells are first found in the medial portion of the lateral mesoderm
CHEN, H. K.	1930	Necturus maculosus	Germ cells arise from the portion of the mesoblastic sheet between the axial and lateral plates
McCOSH, G. K.	1930	Ambystoma maculatum	Germ cells are derived from the lateral mesoderm and the germinal epithelium

Apoda.

SEMON, R.	1891	Ichthyophis glutinosus	Germ cells are derived from the germinal epithelium

Anura.

Author	Year	Species	Results
Goette, A.	1875	Bombinator igneus	Germ cells are first found extra-regionally in the mesoderm
Kolessnikow, N.	1878	Rana esculenta Rana temporaria Bufo cinereus Bufo variabilis	Germ cells are derived from the germinal epithelium
Nussbaum, M.	1880	Rana fusca [1]	Germ cells are first found extra-regionally in the mesoderm
Hoffmann, C. K.	1886	Rana esculenta Rana temporaria Bufo cinereus Alytes obstetricans	Germ cells are derived from the germinal epithelium
Bouin, M.	1900 1901	Rana temporaria	Germ cells are derived from mesenchymal, peritoneal and perhaps entodermal cells
Allen, B. M.	1906	Rana pipiens	Germ cells are first found in the entoderm
Dustin, A. P.	1907	Rana fusca Bufo vulgaris	Germ cells are derived from the medial portion of the lateral mesoderm (gonotome), and from peritoneal cells
King, H. D.	1908	Bufo lentiginosus	Germ cells are first found in the entoderm
Kuschakewitsch S.	1908 1910	Rana esculenta	Germ cells are derived from the entoderm and from peritoneal and „paragonial" elements
Champy, C.	1913	Rana temporaria	Germ cells are derived from the mesoderm mediad to the nephrotome (gonotome)
Witschi, E.	1914	Rana temporaria	Germ cells are first found in the entoderm
Swingle, W. W.	1921 1926	Rana catesbeiana	Primordial germ cells are derived from the entoderm. Secondary cells are derived from the direct descendants of the primordial germ cells
Beccari, N.	1924	Bufo viridis	Germ cells are first entodermal in position and become secondarily attached to the lateral plates of the mesoderm
Humphrey, R. R.	1925	Rana pipiens Rana sylvatica Rana palustris Rana clamitans Bufo americanus [2]	Germ cells are first found in the entoderm
Bounoure, L.	1924a 1929	Rana temporaria Bufo vulgaris	Germ cells are first found in the entoderm
Perle, S.	1927	Bufo vulgaris	Germ cells are first found in the entoderm
Witschi, E.	1929a	Rana sylvatica	Germ cells are first found in the entoderm
Christensen, K.	1930	Rana pipiens	Germ cells are first found in the entoderm

[1] *Rana fusca* is synonymous with *Rana temporaria*.
[2] *Bufo americanus* is synonymous with *Bufo lentiginosus*.

In *Rana cantabrigensis,* there can be no doubt regarding the sexual nature of the primordial germ cells. They are not temporary structures, nor ordinary somatic cells in preparatory stages of mitotic activity, or in abnormal hypertrophy or degeneration. By tracing their future history, they are seen to give rise to the definitive functional sex elements. As already described, the primordial germ cells of *Rana cantabrigensis* antedate the germ glands and have an origin remote from the gonadogenic region. They become first recognizable as distinct structures, at the time when they are separated from the entoderm and become included in the mesenterial anlage, which has meanwhile been formed by the approximation of the lateral plates of the mesoderm. This fact has been demonstrated by ALLEN (1906) on *Rana pipiens,* and by many subsequent investigations on various other species of anurans. GOETTE (1875), NUSSBAUM (1880), and BOUIN (1900, 1901) have followed the germ cells extra-regionally to varying extents in the mesoderm, but have invariably failed to trace them further back to the yolk entoderm. KUSCHAKEWITSCH (1910) claims that in delayed fertilization experiments, the primordial gonocytes do not arise from the entoderm, and ,,*Die erste unpaare Keimanlage bildet sich auf Kosten der medialen Ränder der beiden Seitenplatten, die sich in der Medianlinie vereinigen*" (p. 140). DUSTIN (1907) for *Rana fusca* (= *Rana temporaria*) and *Bufo vulgaris,* and CHAMPY (1913) also for *Rana temporaria,* have denied an entodermal origin of germ cells, to adopt the gonotome theory for amphibians. According to them, the germ-cell anlage in these vertebrates is paired, segmental and mesodermal, instead of being single, medial, entodermal and unsegmented. Their view has not been confirmed, and has repeatedly been rejected by many writers. More recently, BECCARI (1924) has described in *Bufo viridis* an early migration of germ cells from the entoderm to the lateral plates of the mesoderm. Germ cells, in this position, are subsequently carried by the inward growth of the lateral plates, to the median line of the embryo where they constitute the germ cord. A similar condition has been recorded for the Urodela (ABRAMOWICZ, 1913; BOUNOURE, 1925). In our material, however, there is no evidence for this. Examination of many specimens has revealed no elements of germ-cell type in the lateral plates, previous to their medial union to form the anlage of the dorsal mesentery.

While migrating in the mesoderm to their definitive position in the gonad fundament, the germ cells are first in the form of a median germ cord, which is equivalent to the «*ébauche génitale primordiale*» (BOUIN), «*ébauche génitale impaire*» (DUSTIN), or ,,*unpaarem medianem Keimzellstrang*" (KUSCHAKEWITSCH), and are later segregated into paired germ ridges or «*ébauches des glands bilatérales définitives*» in DUSTIN's designation. During these developments, BOUIN (1901), and DUSTIN (1907) have attempted to show a transformation of small somatic cells into germ cells by accession and increase of yolk materials. HUMPHREY (1925) has argued with reason that "Such an increase in size and yolk content of embryonic cells is obviously contrary to the usual course of embryonic differentiation, during which cells ordinarily become smaller and poorer in deutoplasm." KERR (1919, p. 271), however, believes that yolk may be stored secondarily in particular cells or portions of tissues of a developing embryo. KUSCHAKEWITSCH (1910) describes, figures and discusses at length certain elements which he claims represent transitional stages from peritoneal cells or from the "Paragonien" (derivatives of axial mesenchyme and peritoneum ?), to germ cells at stages immediately preceding the formation of germ ridges (tadpoles 6 mm. long and 17 days of age). He believes that such a transformation does not require a renewed deposition of yolk, as the mesodermal cells, during this period, still retain varying amounts of yolk material. This can not be confirmed in our species. Throughout the entire period of migration, the mesodermal cells are much smaller than the germ cells and contain no yolk. These two types of cells are distinctly different form each other, and there are no questionable cells, which might suggest transition stages between them.

The germ ridges appear to develop generally under the influence of the gonocytes. In normal development, somatic tissues do not usually form germ ridges independently of the germ cells, or previously to their migration to this region. This fact controverts the view upheld by GATENBY (1923) that in all "vertebrates the migratory germ cells arrive in the definitive genital ridge a good while after the coelomic (or germinal) epithelium has been established." EIGENMANN's findings (1891, 1896) lead him to conclude that "Sex ridges are formed wherever sex-cells may be located." BACHMANN (1914) has, however, noted that germinal folds may develop in regions where there are no germ cells. From extirpation and transplantation experiments of germinal materials of primordia of *Ambystoma* embryos, HUMPHREY (1927—1929) has concluded that the primordial germ cells in the urodele appear to play a dominant part in the formation and early development of the gonad. Evidences from late fertilization experiments (KUSCHAKEWITSCH among others), and from certain cases of hypogenitalism as previously reported (CHENG, 1930) indicate that in amphibians, germ ridges may, under certain conditions, be formed in the absence of primordial germ cells, and develop into sterile gonads.

Once in the gonad, the germ cells become eventually arranged in a cortical layer, enclosing an inner core of stroma and externally enclosed by a thin peritoneum. This peritoneal lining does not contain nor proliferate any germ cells; hence, it can not be appropriately called the germinal epithelium. On the other hand, evidences indicate that it is the source of origin of follicle cells (BOUIN, KING, et al.).

During the growth of the indifferent gonads, BOUIN, DUSTIN and KUSCHAKEWITSCH have claimed a neoformation of germ cells from peritoneal and mesenchymal elements. In our specimens, germ cells have at this stage begun active multiplication. They retain their characteristic cytomorphic features, and are readily distinguishable, as hitherto, from the surrounding soma cells. The former are never seen to be converted into the latter, whatever the latter might be, peritoneal, stromal or follicular. Likewise, there is no morphological evidence in support of an alleged differentiation of peritoneal or mesenchymal cells or their derivatives, into the germ cells; nor is there any extensive degeneration of germ cells, which would indicate or necessitate such a differentiation. All multiplication of germ cells is by mitotic divisions of the prexisting germ cells, and by this alone.

Our evidence thus far shows definitely that all the germ cells in later stages of development are derived from, and hence genetically related to, those first recognizable as such. They originate extra-regionally from the yolk entoderm, and become secondarily attached to the mesoderm, migrating through it to the gonadogenic region. The origin or formation of germ cells, from an embryological point of view, is, therefore, purely entodermal, without any contributions from elements of mesodermal tissue. Our material also shows clearly that the germ cells do not have any segmental origin, nor any segmented arrangement during subsequent stages of their development. These observations contradict WALDEYER's theory of a germinal epithelium and RÜCKERT's theory of a gonotome, and are essentially in accord with the results obtained by ALLEN, KING, WITSCHI and others on *Bufo* and *Rana*.

While we consider it established that the germ cells of the Anura are of an entodermal origin, evidence has been accumulating, now seemingly conclusive, that in the Urodela the primordial gonocytes are first found in the mesoderm. It has been suggested that in case of urodeles, the germ cells might have migrated from the entoderm into the potential mesoderm before the two layers are separated. Of this, however, there is no indication from actual observations. BOUIN (1901), DUSTIN (1907), and CHAMPY (1913) have advocated a mesodermal origin of germ cells in anurans. Their claim is not substantiated by our evidence. The interpretations of ABRAMOWICZ (1913), and BOUNOURE (1924b, 1925) regarding an entodermal derivation of germ cells of urodeles has been thoroughly discussed and repudiated by HUMPHREY (1925, 1929a). The lastnamed author contends that in tracing back the germ cells of the Anura and Urodela into earlier and earlier periods of development they would ultimately be found to lie "in the territory in which mesoderm and entoderm were yet fused — the undifferentiated material at and immediately anterior to the blastoporic lip." He further asserts that "the primordial germ cells of vertebrates, regardless of the germ layer in which they ultimately appear, may be said to be derived from the germ ring or, in the meroblastic eggs, the equivalent blastodermic margin. Their later position is secondary — the result of the specific or generic pattern of development..." This is a more concrete, and more or less modified expression of the belief that the germ cells are not evolved from, nor do they truly belong to, any germ layer, but are set aside, *ab ovo*, as specific entities. This conception gains valid support from works on certain vertebrates, in which it has been found possible to trace back the germ cells to a period when the definitive germ layers have not yet been formed or are just forming. EIGENMANN (1891) believes that the germ cells of *Micrometrus* might probably be segregated as early as the fifth cleavage. WOODS (1902) claims to have found the primary gonocytes of *Acanthias* in the entoderm before the differentiation of mesoderm, and later in a region of the embryo near the junction of the 3 germ layers. According to DODDS (1910), the germ cells in *Lophius* appear before the entoderm has separated from the mesoderm. BEARD (1900, 1902) states that the germ cells of *Raja* can be definitely traced back to late stages of cleavage, before the formation of any real embryo. The "embryo" is here used in the sense of soma, as distinguished from the germ cells or the germplasm. WHEELER (1899) and OKKELBERG (1921) have been able to recognize the germ cells of cyclostomes in the mesentoderm before the differentiation of definite entodermal and mesodermal tissues. Recently, BOUNOURE (1925, 1927c, 1929) claims to have followed the primordial germ cells of *Rana temporaria* back to the very beginning of gastrulation, when they occupy a definite place in the floor of the blastocoele.

From theories and evidences given above, it becomes probable that in the case of *Rana cantabrigensis*, the entodermal origin of germ cells may not be the real or ultimate origin of these cells. The close similarity of germ cells to entodermal cells does not necessarily imply that the germ cells are actually entodermal cells, or are derived from such. Crucial evidence is likewise lacking that the germ cells as such are specific products of early cleavage, a contention which amounts to claiming an actual morphological continuity of germ cells. It is plausible, however, from a consideration of all available data, that the germ cells are developed precociously before their localization in the entodermal ridge, from an early embryonic or undifferentiated tissue. Such a contention for our species is not purely hypothetical, as BOUNOURE states that he has actually found this to be the case in *Rana temporaria*. The germ cells may, thus, be regarded as continuous with the embryonic materials directly developed from the fertilized ovum. This process of germ-cell formation may probably be analogous with the ordinary procedure of embryonic segregation which leads to the division of labor among the component elements of the body. This view affirms the essential issues of the theory davanced by MORGAN (1891), and EIGENMANN (1896). The above generalization does not, in the least, conflict with or obscure the feature characteristic of the germ cells of *Rana cantabrigensis*, that they, though being formed very early during ontogenetic segregation and specialization, have remained in an apparent primitive condition for a considerable period, and are among the last to undergo histological differentiation. The fact that the germ cells always arise in a perfectly definite way, and that they are genetically specific and continuous at all stages after their first appearance, gives support to our belief that the formation of germ cells in normal embryonic development is an orderly process of the determinate type, specific in time and definitely localized in space.

While it seems evident from the above discussion that germ cells are precociously developed during early ontogeny, nevertheless the question of germ-cell origin cannot be considered entirely settled. Many of the previous investigators have confined their work to the early history of the germ cells and have arrived at some important conclusions, assuming that the primordial gonocytes persist to form definitive sex elements and constitute the only source from which the latter elements are derived. Such an assumption needs substantiation in every case, because evidences have been accumulating that in many species of amphibia there appears in the course of ontogenetic development a second generation of germ cells, which are generally claimed to have been derived from a different source. BOUIN (1901) contends that in *Rana temporaria* the primary gonocytes degenerate in large numbers, and are expelled from the gonad into the body cavity, while a second generation of germ cells arises by transformation of mesenchymal and peritoneal cells. DUSTIN (1907) describes in other amphibians the appearance of a secondary line of gonocytes derived from a peritoneal source. APEHL and POLUS (1912), and ABRAMOWICZ (1913) likewise postulate a similar neoformation of germ cells for the Urodela. KUSCHAKEWITSCH (1910) holds that the primary gonocytes of *Rana* are derived from the entoderm, while the secondary gonocytes are derived from peritoneal, *"paragonial"* and rete-cord elements. According to Gatenby (1916), the first germ cells which migrate from the entoderm are relatively unimportant. "Through the peritoneum, the mesoderm supplies by far the greatest number of germ cells." HARGITT (1924) concludes, from a general review of literature, that the primary germ cells of amphibians mostly degenerate, and the definitive germ cells are derived from a second generation, at least some of which develop from the peritoneal epithelium. SWINGLE (1921) contends that in *Rana catesbeiana*, practically all of the primary germ cells degenerate through an abortive maturation cycle. A second generation of germ cells appears, which, according to his later investigation (1926) are derived from a few lineal descendants of the primordial gonocytes, and are not transformed from mesothelial elements of the gonad. McCOSH (1930b), working on *Ambystoma maculatum*, believes that in this form a few of the functional sex cells may be derived from a small portion of the primordial germ cells, while the majority of them are developed from the germinal epithelium. Evidences showing the development of more than one generation of germ cells have also been obtained from every other group of vertebrates including man (Böhi, 1904; von Winiwarter and Sainmont, 1908; von Winiwarter, 1910; Felix, 1912; von Berenberg-Gossler, 1912; 1914; Firket, 1914, 1920; Kirkham, 1916; Kingery, 1917; Essenberg, 1923; Hargitt, 1925—1929; Stieve, 1927; Butcher, 1928, 1929; Swezy, 1929). A number of the above mentioned authors have agreed that the first generation of germ cells may give rise to some of the definitive sex elements. In case of mammals, however, many regard the primordial germ cells as ordinary or degenerating somatic cells (cf. our previous discussion on the germinal epithelium theory). Some workers think it probable that the primary germ cells in the very process of degeneration exert the necessary stimulation for the formation of secondary gonocytes from the germinal epithelium. This, however, fails to apply to those cases where secondary germ cells are reported to have been developed in the total absence of the primary gonocytes. Von Berenberg-Gossler holds that the so-called primordial germ cells are merely entodermal wandering cells, contributing largely to the formation of the Wolffian duct. Some of them may enter the germ gland, where they become transformed into mesenchymal cells, and then retransformed into germ cells. Firket among others has advocated a phylogenetic interpretation for the regression of the first generation of germ cells, according to which the primary gonocytes may function in lower vertebrates, but are replaced by secondary gonocytes in the higher vertebrates.

In view of the above researches, it becomes obvious that any attempt to ascertain the germ-cell origin in a given species cannot be considered as adequate without thoroughly investigating the entire germ-cell cycle from the embryonic origin of the primordial germ cells to the formation of the definitive functional sex elements, including the origin of the subsequent generations of germ cells after spawning. In our succeeding papers on the later history of germ-cell development in this species, the question of origin or genesis of the germ cells will receive further consideration.

Having discussed thus far the embryological origin of germ cells, we cannot entirely overlook the genetical aspects of this problem, which are centered in the germplasm theory. According to the modern view of Weismann's hypothesis, all cells of the body, including those concerned in reproduction, contain identical amounts of inherited germplasm in the nucleus, somatic differentiation being confined to the cytoplasmic substances of the cell body. If this is true, it is conceivable, then, that any soma cells, in an undifferentiated or dedifferentiated condition, would be capable of transforming into germ cells under proper stimulation and appropriate conditions. That such a transformation actually takes place has been claimed by many, among whom may be mentioned Child (1906) on Moniezia, HARGITT (1913—1919) on Coelenterates, and GATENBY (1916—1924), SIMKINS (1923—1928), HARGITT (1923—1930), and McCOSH (1930) on various vertebrates. Whatever may be the significance of these evidences, the mass of our data shows positively the non-existence of such transformation in the normal embryonic development of *Rana cantabrigensis,* and strongly suggests a genetical integrity and genealogical continuity of germ cells after their formation.

Experimental methods have been devised to solve the disputed question of transformation of somatic cells into germ cells. LEVI (1904—1905) attempted to destroy the gonads of toad larvae *(Bufo vulgaris)* by cauterizing the dorsal part of the coelomic cavity. Many of the operated animals died in the course of a few days. Those which survived later regenerated their gonads, probably on account of incomplete castration. KUSCHAKEWITSCH (1910) working on *Rana esculenta,* found that delayed fertilization of frog eggs inhibited the development of early germ cells from the yolk entoderm. Relatively late in life, secondary gonocytes were evolved by transformation of rete-cord cells, and they gave rise to spermatogonia only. WITSCHI (1914), and BOUNOURE (1927a, b), in their studies of germ-cell development in tadpoles produced from overripe eggs of frogs and toads, found that in these tadpoles, the primordial germ cells were developed from the entoderm, only being more or less modified in their appearance and migrating to the gonad more slowly than normal. WITSCHI (1929a) considered it possible, however, that "in extreme cases, overripeness might completely suppress the migration of entodermal germ cells." HUMPHREY (1927a) attempted to suppress germ-cell migration by the employment of chemical and operative procedures. He stated that after such procedures the genital ridges developed into sterile (or practically sterile) gonads, in which secondary germ cells appeared to develop in older stages. The same author (1927b, c) reported that in *Ambystoma* embryos, complete unilateral extirpation of primordial germ cells and the associated mesoderm resulted in the absence of a typical gonad on the operated side of the body. No indication of a possible formation of secondary gonocytes was found up to 290 days after operation. In footnotes 1 and 3 of the same paper, the author mentioned that he had found regeneration of germ cells in certain animals of another series, upon which a similar operation had been performed at an earlier period of development. In adult tritons, ARON (1924) found no regeneration of testes following complete castration. MORGAN and MACNAB (1926) reported that in *Diemyctylus,* after double ovariectomy, most of the operated animals did not regenerate their ovaries. In some cases, ovarian tissue was found present at autopsy, which was considered as probably due to some compensatory hypertrophy of the bits of tissue unremoved at the time of operation. Similar evidence has been obtained by ADAMS and KIRKWOOD (1928) in their castration experiments on males and females of *Triturus viridescens*. ADAMS (1928, 1930), in her work on ovarian grafting in castrated males of the same species, found no testicular tissue regenerated in the experimental animals. In the chick, REAGAN (1916) removed from young embryos an extra-embryonic area in which the primordial germ cells are supposed to be located. He did not find any typical germ cells in the gonads of such operated embryos even after 5 days of incubation. WILLIER's implantation experiments (1926) give evidence that in the chick embryos no germ cells develop in the absence of the primordial germ cells. DOMM (1927) has performed an extensive series of gonadectomies on hens. His work shows no regeneration of ovarian tissue following complete ovariectomy. In medical literature, there are many records of pregnancy in women after removal of both ovaries (MORRIS, 1901; DORAN, 1902; KYNOCH, 1902; MEREDITH, 1904, et al.). Ovarian regeneration in these cases may be attributed to incomplete extirpation of the germinal tissue. CASTLE and PHILLIPS (1911, 1913) removed gonads from a large number of female guinea pigs and rabbits. Subsequent examination revealed a regeneration of ovarian tissue in a few of the experimental animals. In rats and mice, DAVENPORT (1925), PARKES and BELLERBY (1926), PARKES, FIELDING and BRAMBELL (1927), HANSON and HEYS (1927), and PALLOT (1928) have claimed that, following complete ovariectomy, new ovaries, may, under favorable conditions, be regenerated from a somatic source. On the other hand, HATERIUS (1927, 1928) and more recently PENCHARZ (1929) working on the same animals, have maintained that the ovarian tissue, after having been completely excised, does not arise *de novo* from nongerminal material. Experimental evidence now available is by no means conclusive concerning the genetical continuity and independence of the germ cells, and the existing controversy in the published results shows the need of further investigation, before any generalization can be attempted on this question.

Migration of Germ Cells.

SIMKINS, HARGITT and others, from their work on germ cells of mammals and reptiles, have objected to the theory of extraregional origin of germ cells and their subsequent migration to the gonad. All germ cells, in their opinion, are differentiated from the peritoneal epithelium which first forms and later invests the germ gland. BRAMBELL (1927) even considers it probable that the germ cells of the mouse, originating from the germinal epithelium, might migrate out of the gonad and constitute the extraregional germ cells. In wood-frogs, our evidence is definite and demonstrative of the fact that primordial germ cells do not originate within the gonad, but elsewhere in the body, from whence they later migrate to the gonadogenic region. The fact of germ-cell migration is clearly shown in a complete series of stages. Successive steps of this process are frequently encountered in the very same specimen. Quantitative studies have demonstrated that germ-cell migration is not of a sporadic nature, but a general process occurring in every young tadpole.

In our wood-frogs, and probably in all of the Anura, the germ cells migrate more or less in masses, instead of singly, as in many other forms. Apparently owing to this mode of progression, stray germ cells are relatively of rare occurrence. Only a few such cells have been encountered in any one tadpole which shows this abnormality. In other vertebrates, BEARD (1902), JARVIS (1908), ALLEN (1911), and others have recorded high percentages of abnormally-placed germ cells. Data on germ-cell ectopia are available for toads (KING, 1908; BECCARI, 1924), and for urodeles (Humphrey among others). In frogs, WITSCHI (1914), BOUNOURE (1925), and HUMPHREY (1925) have reported scattered germ cells in malpositions, as in the intestinal wall, in the coelomic epithelium, within the folds of the dorsal mesentery, or somewhere between the mesentery attachment and the genital folds. GATENBY (1916) states that he has found germ cells outside the gonad in frogs long after germ-cell migration, and contends that they are derived *in situ* from retro-peritoneal cells. OBRESHKOVE (1924), and HARGITT (1924) likewise consider the isolated ectopic germ cells as having been developed from the adjacent somatic tissue. This conception is unjustified by histological data, and is definitely disproved by our embryological evidence on the development and migration of germ cells.

In most cases, the abnormally-situated germ cells appear to be belated or left behind in migration. It is conceivable that these cells, if closely adjacent to the germ gland, may in time move into it. If not, they would eventually degenerate *in situ*, but might persist and develop, under favorable stimulation and condition, into germinal masses in abnormal regions. These neoplastic masses may be included in certain tissues or organs of the body, or may be free, forming independent germinal structures, which subsequently give rise to gonadic multiplicity or heterotopy, as recorded by NUSSBAUM (1906), GERHARTZ (1906), and DAUVART (1926, 1927) in adult frogs. EIGENMANN (1896) reported that in a Cymatogaster embryo, a few of the germ cells were stranded in their migration and formed an extra germ ridge independently of and separated from the germ ridges occuring in the normal position. WITSCHI (1914) found in a newly metamorphosed frog *(Rana temporaria)*, besides 2 normal gonads, an ,,akzessorische Keimdrüse" equally well developed. CAROLI (1926, 1927) observed in the left kidney of a sexually mature male toad *(Bufo viridis)*, a rounded formation of about 200 micra in diameter, which was composed of male germ cells in a healthy condition. In his opinion, this was due to an actual inclusion of a portion of the genital crest into the renal parenchyma during the embryonic development. But inferred from our data on germ-cell ectopia, this anomaly may best be explained as an instance of abnormal migration of germ cells. The fact that we have found aberrant germ cells actually within the renal blastema renders this explanation even more probable. According to BEARD (1902c), some of the ectopic primordial gonocytes might survive to give rise to dermoid cysts, or the so-called embryomas.

The path of germ-cell migration is determinate and definite. The germ cells begin migration as a median mass from the entoderm. Inferred from the initial position of germ cells in most of the lower vertebrates so far studied, it is possible, as suggested by ALLEN (1906), that the germ cells of the Anura may migrate precociously from more lateral portions of the body to the median entodermal ridge. The recent findings of BECCARI (1924) on *Bufo viridis,* and BOUNOURE (1925, 1927c) on *Rana temporaria* lend support to this assumption. The latter author claims to have actually traced the germ cells to the bilateral, symmetrically-placed foci in the entoderm.

Eventually as the germ cells are separated from the entoderm, they come to lie first in the splanchnic mesoderm, then in the mesoderm of the dorsal mesentery, and from there they migrate to the gonadogenic region on each side. In the Urodela, the migration path is strikingly different. The germ-cell anlage in the lateral mesoderm is originally paired. The paired cell groups, in most forms, approach each other in the midline, but, according to HUMPHREY (1925, 1929a), they never fuse to form a true median structure, such as in the Anura. They eventually separate with the formation of the dorsal mesentery, and are shifted laterally to form the genital ridges.

The process of migration is gradual and is coordinated with the development of the adjoining tissues and structures. It is invariably more advanced anteriorly than posteriorly.

As regards the means of migration, ALLEN (1907) on *Rana pipiens,* advocates the view of an active and independent movement on the part of the germ cells. In many forms, germ cells are seen to assume, during a part of their early history, slightly irregular outlines, which condition has been interpreted as indicative of their ameboid mode of progression. ALLEN (1906) suggests that the germ cells of reptiles, despite their rounded form, might undergo slight ameboid movements brought about by imperceptible changes in their shape. WOODGER (1925) has described and figured unmistakable pseudopodia in the gonocytes of the fowl. SWIFT (1915), working on the same animal, shows that the distribution of primordial germ cells is invariably in favor of the left gonad, and regards this as a positive proof of their active migration. Some authors have called attention to the parallelism between the independent motility of germ cells and that of the nervous and muscular elements during embryonic development. HUMPHREY (1925), on the other hand, supports the view that germ-cell migration in amphibians, is a passive process brought about by the growth shiftings and readjustments of the related structures. A similar interpretation has been advanced for fishes (RICHARDS and THOMPSON, 1921; and HANN, 1927) and for reptiles (VON BERENBERG-GOSSLER, 1914). It has also been proposed that germ cells may be carried along by ameboid movements of their follicle cells (OKKELBERG, 1921). In general, our evidence is in accord with HUMPHREY's view, but admits the possibility that while in the entoderm, germ cells might have undergone a process of active segregation to reach the appropriate position in the entodermal ridge. Recent work of BOUNOURE on *Rana temporaria* seems to show that such a process actually takes place. The lateral movement of germ cells from the median germ cord to form the paired germ ridges takes place, according to HUMPHREY, "in conjunction with a similar movement of associated parts." In *Rana cantabrigensis*, our evidence is contrary to this, and indicates that this process is coincident with and may possibly be brought about by the approach of the revehent veins medially to form the future post cava (fig. 4). BOUNOURE (1925) states that in *Rana temporaria*, the development of germ ridges takes place during the formation of the postcava. In our preparations, as seen in the figure already referred to, the former process always precedes the latter.

The rare occurrence of scattered germ cells in the mesonephros and in regions far away from the normal migration track may perhaps be explained best on the basis of chance wanderings on the part of these cells Such an explanation necessitates the assumption that germ cells, no matter how large and inert, may be capable of some independent movement, however slight it might be. As to the causes that prompt this active movement and the actual means by which it is accomplished, morphological data are too meager for any suggestions. The view that germ cells may take an active part during certain phases of their migration, and a passive part in others, has been advocated for other vertebrates (WOODS, 1902; DODDS, 1910; JORDAN, 1917; VANNEMAN, 1917; OKKELBERG, 1921; and CHEN 1930). Intravascular migration of germ cells does not normally occur in the frog, as has been demonstrated for birds (VON BERENBERG-GOSSLER, FIRKET, SWIFT, WOODGER, and GOLDSMITH).

Characteristics of Germ Cells.

In *Rana cantabrigensis*, the primordial germ cells are best characterized during early stages by their position or location. They can be readily distinguished from all surrounding mesodermal elements by their wealth of yolk materials. In this respect, however, they resemble the primitive entodermal cells, and this resemblance is apparently due to the late differentiation common to both types of cells. As development proceeds, the entodermal cells soon use up the contained yolk, while the germ cells remain yolk-laden for a considerably longer period. The persistence of yolk affords, then, a precise and a most conspicuous means of identifying the germ cells.

Another characteristic of the germ cells is the absence of mitosis for a long time subsequent to their first appearance. This has been ascribed to the presence of large quantities of yolk in them. Owing to the crowded condition of yolk particles, it is beyond our available means to detect mitotic figures, should they exist as they might. Occasional mitoses during germcell migration have been recorded by SWIFT (1914) in birds, JORDAN (1917) in reptiles, BEARD (1902c) and BACHMANN (1914) in fishes, and DUSTIN (1907) and KUSCHAKEWITSCH (1910) in amphibians. The last-named author, from counts of germ cells in a number of specimens *(Rana esculenta)*, holds that these cells begin to multiply even before they reach the gonadic region. WITSCHI (1914) has re-analyzed his statistical data and points out that they do not support this contention but rather indicate individual variability with respect to germ-cell number. Statistical studies of BÖHI (1904), EIGENMANN (1896), BECCARI (1924), and BOUNOURE (1925) may suffice to show that there is no numerical increase of germ cells during the time of their migration, which period may, thus, be designated as a period of rest.

After the absorption of yolk, the germ cells become noticeable as large, rounded structures. Their nuclei enlarge and soon assume a characteristically polymorphic appearance, which is a surest criterion for the recognition of these cells.

BOUNOURE (1927c, 1929a) has recently reported the presence of a *"chondriome"* in the primary gonocyte of *Rana temporaria,* which is «une masse cytoplasmique juxtanucléaire, d'aspect granuleux et très dense ...» By means of this cytological structure, he claims that he is able to distinguish between the germ cells and other cells of the body in the very early stages when gastrulation has just begun. This structure is of considerable interest, and is unique in the mass of literature on the early embryology of germ cells of frogs.

In certain vertebrates, visible substances apparently similar in nature to the so-called Keimbahn-determinants in the invertebrates have been found (DODDS, 1910; RUBASCHKIN, 1910b, 1912; TSCHASCHIN, 1910; and others). Most of these are still in dispute, and, in no case, has the evidence been completely established. According to HEGNER (1914), "A large number of Keimbahn-determinants which have been described are supposed to consist of nutritive substances," which are likely akin to the yolk materials so abundant in the early germ cells of frogs.

The long retention of yolk by the germ cells, the absence of cell division among them, their large size, and other peculiarities of their cellular structures, all indicate that these cells have been endowed with certain specific properties, and are preserved in an embryonic, or a slightly differentiated condition for a considerable period in their early development. This characteristic feature of the germ cells is rendered more evident and more significant in contrast to the rapid and pronounced development and specialization exhibited by the early somatic tissues.

Rete Cords.

The structures, designated as rete cords in this paper, have been previously described as « cordons médullaires » by Bouin, or as ,,Geschlechtsstränge", ,,Genitalstränge", or ,,Sexualstränge (sex-cords)" by students of the Hertwig school. Swingle (1921) points out that the term "sex cord" is a misnomer, but he as well as many others have consistently used the term apparently as synonymous with "rete cord". Witschi (1929) has definitely adopted the latter term as more appropriate than the former one. Regarding the origin of rete cords, or rather the cells of which these cords are composed, it has been contended that they are derived from the peritoneal epithelium, as for instance by King (1908) in toads, and Abramowicz (1913) in tritons. Recent investigations on both the Anura and Urodela have shown that the rete cord materials originate from the nephrogenous tissue, whence they migrate into the developing gonad. Semon (1891) has noticed in *Ichthyophis,* cellular cords arising from the nephrotome and passing into the germ gland. In other vertebrates, rete cords have been variously postulated or described as being derived from the Wolffian duct (Waldeyer, 1870), from the renal corpuscles (Mihálkowics, 1885; Hoffmann, 1889; Kingery, 1917), from the germinal epithelium (Janošík, 1885; Simkins, 1923; Wilson, 1926), from the stromal mesenchyme (Firket, 1914; Swift, 1915), or from "funnel-cords" together with evaginations from the corpuscles of Bowman (Allen, 1905).

It should be made clear that in wood-frog tadpoles, the cells that are later concerned in the formation of rete cords do not migrate into the germ ridges in a tubular or any other definite form. Even inside the ridge, they do not show any organization, being widely dispersed among the germ cells and their associated small mesenchymal elements. Hence, during this period, there are no rete cords to speak of. Not until the time of morphological sex differentiation do the stroma cells of the gonad, undergo a process of condensation to form the segmentally arranged rete cords. Thus considered, these cords are really a product of stromal condensation. Bouin (1901) has made a similar observation of *Rana temporaria,* in which species the rete cords are differentiated « aux dépens de la zone centrale des tissu mésenchymaeux » of the gonad.

The segmental arrangement of rete cords produces a metameric appearance of the gonad, which condition Witschi (1929a) designated as gonomery, an unsatisfactory term for this usage, since it has been generally accepted to mean the independence of the maternal and paternal chromosomes and nuclei in the products of a zygote (WILSON, 1925; SHARP, 1926).

The rete cords eventually give rise to rete testis in case of the male, and rete ovarii in case of the female. In either case, they constitute the intragonadal communicating network, which becomes functional and essential in the testis, while rudimentary and superfluous in the ovary. WITSCHI (1929b) maintains that from rete cords are derived not only rete tubules, but also seminiferous tubules which have generally been believed to differentiate from sex cords. Our data on this may be best considered with reference to, and in connection with, the sex differentiation of the gonad, which will be described and discussed in a separate publication.

Hypogenitalism.

Owing to its possible bearing on the problem of germ-cell origin, hypogenitalism may be briefly considered here. As the term implies, this condition may be defined as atrophia or underdevelopment of gonads or genital structures. As such it is distinctly an embryological abnormality and cannot be considered as a normal feature of individual variability.

Cases of underdeveloped or atrophied gonads have been encountered sporadically at different stages of larval development, during the time of metamorphosis, and even in the adult frog. A full description and discussion of this subject has been presented in a previous publication (CHENG, 1930). The gonads described contain scanty or no germinal elements, and in most cases, they appear to have undergone histological changes of sex differentiation despite the abnormal condition of their germinal contents. A new case has recently been found (tadpole 63 days old and 42 mm. total length). The specimen shows a well-developed left testis, measuring approximately 0,35 mm in length with a fat-body in front and a short postgonad behind. Figure 12 shows this gland through its greatest diameter, with the left testis for comparison. Structurally the abnormal gonad is packed full of mesenchymal cells with no definite organization and shows no features which are considered characteristic of the ovary or testis. No germinal tissue is found to be present, nor is there any identifiable trace of such. The fat-body also has not developed on this side.

Such abnormal gonads as the one above described can hardly be regarded as having been induced by starvation, nor as being the result of a process of "fatty degeneration" in the interpretation of MARSHALL (1884), or sex reversal as claimed by WITSCHI among others. For a review and discussion of literature on this subject, reference should be made to our previous paper already quoted. We consider it probable that hypogenitalism is primarily an abnormality in germ-cell development, which may be accompanied or followed by other changes in the gonad, and which is not necessarily connected with the question of sex differentiation or transformation. Cases of hypogenitalism hitherto encountered in *Rana cantabrigensis* favor our conception of a precocious differentiation of germ cells. In case the germ cells fail to form, or to migrate to the gonadogenic region, or to develop properly in the gonad, there is no further differentiation of them from somatic sources at any subsequent stage. Consequently, a condition of hypogenitalism is destined to appear.

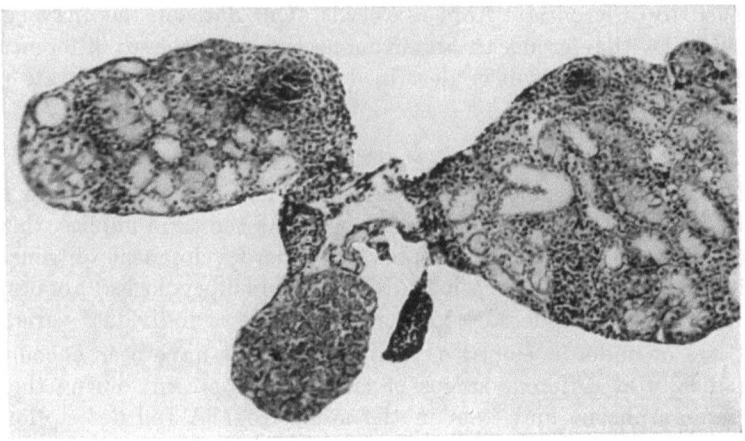

Fig. 12. Tadpole 63 days old and 42 mm. long. Photograph shows a well-developed testis on the left side of the body and an atrophied gonad on the right. The material was sectioned from behind forward, so that the right side of the illustration corresponds to the right side of the reader, and the left to left.

The occurrence of this abnormality can, thus, be adequately explained by, and is, in many respects, indicative of an early segregation and genetical specificity of germ cells. Whether hypogenitalism is a permanent condition, or merely a transient feature during a certain period of development needs further investigation. Apparently, it may be either, depending upon the structural and developmental condition of the gonad in question.

Summary.

1. The primordial germ cells of *Rana cantabrigensis* are first associated with the entoderm during early ontogenesis. They become readily recognizable as they come to lie in the mesoderm.

2. While migrating in the mesoderm, the germ cells are first in the form of a median germ cord. They become subsequently shifted laterally to form paired germ ridges, one on each side of the body, closely adjacent to the dorsal attachment of the gut mesentery.

3. The early germ cells are characterized by their location, and by their undifferentiated or slightly differentiated condition.

4. The germ cells migrate *en masse*. Their migration is mainly passive in nature. Evidences are presented which indicate that they may be capable of a slight independent motility.

5. Ectopic germ cells have been encountered sporadically. They may degenerate early, or persist to form accessory germinal structures in subsequent stages of development.

6. The germ ridges gradually grow in size and develop into the primordial germ glands, each of which is divisible into 3 portions: Progonad, gonad proper or definitive germ gland, and postgonad.

7. The stromal tissue of the germ gland is mainly composed of cells migrating from mesonephric blastema. Rete cords are formed by a process of condensation of the stroma cells.

8. The definitive germ gland takes on the character of an indifferent gonad characterized by a cortical germinal layer, and by the presence of a primary genital cavity and segmentally arranged rete cords in the medullary region.

9. Germ cells in the indifferent gonad soon begin to multiply. They are genetically continuous with the primordial germ cells, and are not differentiated from any somatic source.

10. Evidence from hypogenitalism favors our conception of a precocious segregation and genetical continuity of germ cells.

Literature cited.

Abramowicz, H.: Die Entwicklung der Gonadenanlage und Entstehung der Gonocyten bei Triton taeniatus (Schneid). Morph. Jb. 47, 593—644 (1913). — **Adams, A. E.:** The production of hypertrophy of the Müllerian ducts in castrated males of Triturus viridescens by ovarian grafts. Anat. Rec. 41, 26—27 (1928) (Abstract). — Studies on sexual conditions in Triturus viridescens. The effects of ovarian grafts in castrated males. J. of exper. Zool. 55, 63—86 (1930). — **Adams, A. E. and Kirkwood, E. S.:** The effects of gonadectomy in Triturus viridescens. Anat. Rec. 41, 35 (1928) (Abstract). — **Allen, B. M.:** The embryonic development of the ovary and testis of the mammals. Amer. J. Anat. 3, 89—154 (1904). — The embryonic development of the rete-cords and sex-cords of Chrysemys. Amer. J. Anat. 5, 79—94 (1905). — The origin of the sex-cells of Chrysemys. Anat. Anz. 29, 217—236 (1906). — An important period of the history of the sex-cells of Rana pipiens. Anat. Anz. 31, 339—347 (1907). — Origin of the sex cells in Amia and Lepidosteus. Anat. Rec. 3, 229—232 (1909). — The origin of the sex-cells of Amia and Lepidosteus. J. of Morph. 22, 1—36 (1911a). — The origin of the sex-cells in Necturus. Science N. Y. 33, 268—269 (1911b). — **Allen, E.:** Ovogenesis in the sexually mature mouse. Anat. Rec. 21, 44 (1921) (Abstract). — Oestrous cycle in the mouse. Amer. J. Anat. 30, 297—372 (1922). — Ovogenesis during sexual maturity. Anat. Rec. 25, 116—117 (1923) (Abstract); Amer. J. Anat. 31, 439—482 (1923). — **Amma, K.:** Über die Differenzierung der Keimbahnzellen bei den Copepoden. Arch. Zellforsch. **6**, 497—576 (1911). — **Arai, H.**: Postnatal development of the ovary, with especial reference to number of ova (albino rat). Amer. J. Anat. **27**, 404—462 (1920). **Aron, M.**: Recherches morphologiques et expérimentales sur le déterminisme des caractères sexuels mâles chez les Urodèles. Archives de Biol. **34**, 1—166 (1924). — **Bachmann, F. M.**: The migration of the germ cells in Ameiurus nebulosus. Biol. Bull. Mar. biol. Labor. Wood's Hole **26**, 351—363 (1914). — **Beard, J.**: The morphological continuity of the germ-cells in Raja batis. Anat. Anz. **18**, 465—485 (1900). — The germ-cells of Pristiurus. Anat. Anz. **21**, 50—61 (1902a). — The numerical law of the germ-cells. Anat. Anz. **21**, 189—200 (1902b). — The germ-cells. Part I Raja batis. Zool. Jb. **16**, 615—702 (1902c). — **Beccari, N.**: Studî sulla prima origine delle cellule genitali nei Vertebrati. I. Storia della indagini e stato attuale della questione. Arch. ital. Anat. **18**, 157—226 (1921). — Studî sulla prima origine delle cellule genitali nei Vertebrati. II. Ricerche nella Salamandrina perspicillata. Arch. ital. Anat. **18**, Suppl. 29—95 (1922). — Studî sulla prima origine delle cellule genitali nei Vertebrati. III. Ricerche nel Bufo viridis. Arch. ital. Anat. **21**, 332—374 (1924). — **Berenberg-Gossler, H. von**: Die Urgeschlechtszellen des Hühnerembryos am 3. und 4. Bebrütungstage, mit besonderer Berücksichtigung der Kern- und Plasmastrukturen Arch. mikrosk. Anat. **81**, 24—72 (1912). — Über Herkunft und Wesen der sogenannten primären Urgeschlechtszellen der Amnioten. Anat. Anz. **47**, 241—264 (1914). — **Böhi, U.**: Beiträge zur Entwicklungsgeschichte der Leibeshöhle und der Genitalanlage bei den Salmoniden. Morph. Jb. **32**, 505—586 (1904). — **Bouin, M.**: Ébauche génitale primordiale chez Rana temporaria (L.). Note preliminaire. Bibl. Anat. **8**, 103—108 (1900). — Histogenèse de la glande génitale femelle chez Rana temporaria (L.). Archives de Biol. **17**, 201—381 (1901). — **Boulenger, G. A.**: A monograph of the American frogs of the genus Rana. Proc. amer. Acad. Arts a. Sci. **55**, 413—480 (1920). — **Bounoure, L.**: Dérivés endodermiques dorsaux et premiére ébauche génitale chez les batraciens anoures. C. r. Acad. Sci. Paris **178**, 339—341 (1924a). — Origine des gonocytes primaires chez les Urodèles et signification des ces éléments chez les amphibiens en général. C. r. Acad. Sci. Paris **179**, 1082—1084 (1924b). — L'origine des gonocytes et l'évolution de la première ébauche génitale chez les batraciens. Ann. Sci. Nat. Zool., X. s. **8**, 201—278 (1925). — La surmaturation ovulaire influt-elle sur l'origine des gonocytes primaires chez Rana temporaria L.? C. r. Acad. Sci. Paris **184**, 401—403 (1927a). — Les gonocytes primaires chez les embryons des crapauds issues d'oeufs soumis a une surmaturation utérine. C. r. Acad. Sci. Paris **184**, 549—551 (1927b). — Le chondriome des gonocytes primaires chez Rana temporaria et la recherche des éléments génitaux aux jeunes stades du développement. C. r. Acad. Sci. Paris **185**, 1304 (1927c). — Sur un caractère cytologique essentiel des gonocytes primaires chez la grenouille rousse. C. r. Soc. Biol. Paris **101**, 703 (1929a). — Sur l'existence des gonocytes primaires dans l'embryon de la grenouille rousse a partir du début de la gastrulation localisation et migration de ces gonocytes aux différents stades. C. r. Soc. Biol. Paris **101**, 706 (1929b). — **Brambell, F. W. R.**: The development and morphology of the gonads of the mouse. Part I. The morphogenesis of the indifferent gonad and of the ovary. Proc. roy. Soc. B **101**, 391—409 (1927). — **Buchner, P.**: Keimbahn und Ovogenese von Sagitta. Anat. Anz. **35**, 433—443 (1910). — **Burns, R. K.**, jr.: The sex of parabiotic twins in Amphibia. J. of exper. Zool. **42**, 31—90 (1925). — **Butcher, E. O.**: The origin of the definitive ova in the white rat (Mus norvegicus albinus). Anat. Rec. **37**, 13—30 (1927). — Germ-cell origin in the lake lamprey (Petromyzon) marinus unicolor. Anat. Rec. **41**, 78 (1928) (Abstract). — The origin of the germ cells in the lake lamprey (Petromyzon marinus unicolor). Biol. Bull. Mar. Labor. Wood's Hole **56**, 87—100 (1929). **Caroli, A.**: Su un reperto di elementi germinati nel parenchima renale di Bufo viridis, Laur. Atti. Accad. Fisiocritici Siena X. s. **1**, 509—514 (1926). — Keimelemente im Nieren-Parenchym von Bufo viridis, Laur. Anat. Anz. **63**, 219—223 (1927). — **Castle, W. E. and Phillips, J. C.**: On germinal transplantation in vertebrates. Carnegie Inst. Wash. Publ. **1911**, 144. — Further experiments on ovarian transplantation in guinea pigs. Science N. Y. **38**, 738—786 (1913). — **Champy, C.**: Recherches sur la spermatogénèse des batraciens et les éléments accessoires du testicule. Arch. de Zool. **52**, 13—304 (1913). — **Chen, H. K.**: (Mrs.) Origin of germ cells in Necturus. Shanghai J. Sci. **3**, 44—75 (1930). — **Cheng, T. H.**: Hypogenitalism in Rana cantabrigensis. Papers Mich. Acad. Sci., Arts a. Letters **11**, 369—380 (1930). **Child, C. M.**: The development of germ cells from differentiated somatic cells in Moniezia. Anat. Anz. **29**, 592—597 (1906). — **Christensen, K.**: Sex differentiation and development of oviduct in Rana pipiens. Amer. J. Anat. **45**, 159—187 (1930). — **Coperthwaite, M. H.**: Observations on pre- and postpubertal oogenesis in the white rat, Mus norvegicus albinus. Amer. J. Anat. **36**, 69—89 (1925). — **Dauvart, A.**: Ein Fall von Hodenheterotopie bei Rana temporaria. Roux' Arch. **108**, 138—145 (1926). — Sur un cas d'hétérotopie testiculaire chez la grenouille. C. r. Soc. Biol. Paris **97**, 256—257 (1927). — **Davenport, C. B.**: Regeneration of ovaries in mice. J. of exper. Zool. **42**, 1—12 (1925). — **Dickerson, M. C.**: The frog book. New York: Doubleday, Page & Co. 1906. — **Dodds, G. S.**: Segregation of the germ cells of the teleost, Lophius. J. of Morph. **21**, 563—612 (1910). — **Domm, L. V.**: New experiments on ovariotomy and the problem of sex inversion in the fowl. J. of exper. Zool. **48**, 31—173 (1927). — **Doran, A.**: Pregnancy after removal of both ovaries for cystic tumour. J. Obstetr. **2**, 1—10 (1902). — **Dustin, A. P.**: Recherches sur l'origine des gonocytes chez les amphibiens. Arch. Biol. **23**, 411—522 (1907). — **Eigenmann, C. H.**: On the precocious segregation of the sex-cells in Micrometrus aggregatus Gibbons. J. of Morph. **5**, 481—492 (1891). — The history of the sex cells from the time of segregation to sexual differentiation in Cymatogaster. Amer. microsc. Soc. **17**, 172—173 (1895). — Sex-differentiation in the viviparous teleost Cymatogaster. Arch. Entw.mechan. **4**, 125—179 (1896a). — The bearing of the origin and differentiation of the sex cells in Cymatogaster on the idea of the continuity of the germ plasm. Amer. Naturalist **30**, 265—271 (1896b). — **Essenberg, J. M.**: Sex-differentiation in the viviparous teleost Xiphophrous helleri Heckel. Biol. Bull. Mar. Labor. biol. Wood's Hole **45**, 46—97 (1923). — **Fedorow, V.**: Über die Wanderung der Genitalzellen bei Salmo fario. Anat. Anz. **31**, 219—223 (1907). — **Felix, W.**: The development of the urogenital organs. II. The development of the reproductive glands and their ducts. Manual of Human Embryology by F. Keibel, and F. P. Mall, Vol. 2, p. 881—979. 1912. — **Felix, W. u. Bühler, A.**: Die Entwicklung der Keimdrüsen und ihrer Ausführgänge. Handbuch der vergleichenden und experimentellen Entwicklungslehre der Wirbeltiere von O. Hertwig, Bd. 3, S. 619—896. 1906. — **Firket, J.**: Recherches sur l'organogénèse des glandes sexuelles chez les oiseaux. Archives de Biol. **29**, 201—351 (1914); **30**, 393—516 (1920a). — On the origin of germ cells in higher vertebrates. Anat. Rec. **18**, 309—316 (1920b). **Foley, J. O.**: Observations on germ-cell origin in the adult male teleost, Umbra limi Anat. Rec. **35**, 11—12 (1927a) (Abstract); **35**, 379—399 (1927b). — **Fürbringer, M.**: Zur vergleichenden Anatomie und Entwicklungsgeschichte der Exkretionsorgane der Vertebraten. Morph. Jb. **4**, 1—111 (1878). — **Fuss, A.**: Über extraregionäre Geschlechtszellen bei einem menschlichen Embryo von vier Wochen. Anat. Anz. **39**, 407—409 (1911). — Über die Geschlechtszellen des Menschen und der Säugetiere. Arch. mikrosk. Anat. **81**, 1—23 (1912). — **Gasparoo, E.**: Osservazioni sull' origine delle cellule sessuali nel Gongylus ocellatus. Monit. zool. ital. **19**, 105—116 (1908). — **Gatenby, J. B.**: The transition of peritoneal epithelial cells into germ cells in some Amphibia Anura, especially in Rana temporaria. Quart. J. microsc. Sci. **61**, 275—300 (1916). — Further evidence on the transition of peritoneal cells into germ cells in Amphibia. J. roy. microsc. Soc. **1923**, 409—416 (1923). — The transition of peritoneal epithelial cells into germ-cells in Gallus bankiva. Quart. J. microsc. Sci. **68**, 1—16 (1924). — **Gerhartz, H.**: Multiplizität von Hoden und Leber. Anat. Anz. **28**, 522—528 (1906). — **Goette, A.**: Die Entwicklungsgeschichte der

Unke (Bombinator igneus). Leipzig: L. Voß 1875. — **Goldsmith, J. B.**: The history of the germ cells in the domestic fowl. J. Morph. a. Physiol. **46**, 275—316 (1928). — **Hall, R.W.**: The development of the mesonephros and the Müllerian ducts in Amphibia. Bull. Mus. comp. Zool. Harvard Coll. **45**, 31—126 (1904). — **Hann, H. W.**: The history of the germ cells of Cottus bairdii Girard. J. Morph. a. Physiol. **43**, 427—498 (1927). **Hanson, F. B.** and **Heys, F.**: On ovarian regeneration in the albino rat. Proc. Soc. exper. Biol. a. Med. **25**, 184—185 (1927). — **Hargitt, G. T.**: Germ cells of Coelenterates. I. Campanularia flexuosa. J. of Morph. **24**, 383—420 (1913). — Germ cells of Coelenterates. II. Clava leptostyla. J. of Morph. **27**, 85—98. (1916). — Germ cells of Coelenterates. III. Aglanthia digitalis. IV. Hybocadon prolifer. J. of Morph. **28**, 593—642 (1917). — Germ cells of Coelenterates. V. Eudendrium ramosum. J. of Morph. **31**, 1—24 (1918). — Germ cells of Coelenterates. VI. General considerations, discussion, conclusions. J. of Morph. **33**, 1—60 (1919). — The seasonal production of new spermatogonia in the adult salamander, Diemyctylus. Anat. Rec. **26**, 338—339 (1923) (Abstract). — Germ-cell origin in the adult salamander, Diemyctylus viridescens. J. Morph. a. Physiol. **39**, 63—112 (1924). — Primordial sex-cells of the albino rat. Anat. Rec. **29**, 107 (1924) (Abstract). — The formation of the sex glands and germ cells of mammals. I. The origin of the germ cells in the albino rat. J. Morph. a. Physiol. **40**, 517—558 (1925). — The formation of the sex glands and germ cells of mammals. II. The history of the male germ cells in the albino rat. J. Morph. a. Physiol. **42**, 253—306 (1926). The formation of the sex glands and germ cells of mammals. III. The history of the female germ cells in the albino rat to the time of maturity. Wistar Inst. Bibl. Service **1929**, Nr 131. — The formation of the sex glands and germ cells of mammals. IV. Continuous origin and degeneration of germ cells in the female albino rat. Wistar Inst. Bibl. Service **1930** a, Nr. 137. — The formation of the sex glands and germ cells of mammals. V. Germ cells in the ovaries of adult, pregnant, and senile albino rats. Wistar Inst. Bibl. Service **1930** b, Nr 148. — **Haterius, H. O.**: An experimenta study of ovarian regeneration in mice. Proc. Soc. exper. Biol. a. Med. **24**, 784—786 (1927); Physiologic. Zool. **1**, 45—54 (1928). — **Hegner, R. W.**: Effects of removing the germ-cell determinants from the eggs of some Chrysomelid beetles, preliminary report. Biol. Bull. Mar. Labor. biol. Wood's Hole **16**, 19—26 (1908). — The origin and early history of the germ-cells in some Chrysomelid beetles. J. of Morph. **20**, 231—296 (1909). — Studies on germ cells I. The history of the germ cells in insects with special reference to the Keimbahndeterminants in animals. II. The origin and significance of the Keimbahn-determinants in animals. J. of Morph. **25**, 375—510 (1914a) — The germ-cell cycle in animals. Macmillan Company, New York 1914b. — **Hoffmann, C. K.**: Zur Entwicklungsgeschichte der Urogenitalorgane bei den Anamnia. Z. Zool. **44**, 570—643 (1886). — Zur Entwicklungsgeschichte der Urogenitalorgane bei den Reptilien. Z. Zool. **48**, 260—300 (1889). — Étude sur le développement de l'appareil urogenital des oiseaux. Verhl. Akad. Wetensch. Amsterd., Wis.en natuurkd. Afd., II. s. **1**, Nr 4, 54 (1893). — **Howe, R. H.,** jr.: North American wood frogs. Proc. Boston Soc. Nat. Hist. **28**, 369—374 (1899). — **Humphrey, R. R.**: The primordial germ cells of Hemidactylium and other Amphibia. J. Morph. a. Physiol. **41**, 1—43 (1925). — Modification or suppression of the so-called migration of primordial germ cells in anuran embryos. Anat. Rec. **35**, 41 (1927a) (Abstract). — Unilateral extirpation of the primordial germ cells in Amblystoma; Its effect upon the development of the gonad. Anat. Rec. **35**, 15 (1927b) (Abstract); J. of exper. Zool. **49**, 363—400 (1927c). — The fate of the primordial germ cells of Amblystoma in grafts implanted in the somatopleure of other embryos. Anat. Rec. **35**, 40—41 (1927d) (Abstract). — The differentiation of the intermediate mesoderm of Amblystoma in grafts implanted in the somatopleure of other embryos. Anat. Rec. **38**, 48—49 (1928a) (Abstract). — The developmental potencies of the intermediate mesoderm of Amblystoma when transplanted into ventrolateral sites in other embryos: The primordial germ cells of such grafts and their rôle in the development of a gonad. Anat. Rec. **40**, 67—102 (1928b). — Sex differentiation in gonads developed from transplants of the intermediate mesoderm of Amblystoma. Biol. Bull. Mar. Labor. biol. Wood's Hole **55**, 317—339 (1928c). — The early position of the primordial germ cells in urodeles; Evidence from experimental studies. Anat. Rec. **42**, 301—314 (1929a). — Studies on sex-reversal in Amblystoma. II. Sex differentiation and modification following orthotopic implantation of a gonadic preprimordium. J. of exper. Zool. **53**, 171—221 (1929b). — **Janosik, J.:** Histologisch-embryologische Untersuchungen über das Urogenitalsystem. Sitzgsber. Akad. Wiss. Wien, Math.-naturwiss. Kl. **91**, 97—199 (1885). — **Jarvis, M. M.:** The segregation of the germ cells of Phrynosoma cornutum: Preliminary note. Biol. Bull. Mar. Labor. biol. Wood's Hole **15**, 119—126 (1908). — **Jordan, H. E.:** Embryonic history of the germ cells of the loggerhead turtle (Caretta caretta). Carnegie Inst. Washington, Publ. **251**, 313—344 (1917). — The history of the primordial germ cells in the loggerhead turtle embryo. Proc. nat. Acad. Sci. U.S.A. **3**, 271—275 (1917). — **Kerr, J. G.:** Text-book of embryology, Vol. 2 Vertebrata with the exception of Mammalia. London: Macmillan and Co. 1919. — **King, H. D.:** The oogenesis of Bufo lentiginosus. J. of Morph. **19**, 369—438 (1908). — **Kingery, H. M.:** Oogenesis in the white mouse. J. of Morph. **30**, 261—316 (1917). — **Kingsbury, B. F.:** The morphogenesis of the mammalian ovary: Felis domestica. Amer. J. Anat. **15**, 345—379 (1913). — Interstitial cells of the mammalian ovary. Amer. J. Anat. **16**, 59—96 (1914). — **Kirkham, W. B.:** The germ cell cycle in the mouse. Anat. Rec. **10**, 217—219 (1919) (Abstract). — **Kohno, S.:** Zur Kenntnis der Keimbahn des Menschen. Arch. Gynäk. **126**, 310—326 (1925). — **Kolessnikow, N.:** Über die Eientwicklung bei Batrachiern und Knochenfischen. Arch. mikrosk. Anat. **15**, 382—414 (1878). — **Kuschakewitsch, S.:** Über den Ursprung der Urgeschlechtszellen bei Rana esculenta. Sitzgsber. bayer. Akad. Wiss. Math.-physik. Kl. **38**, 89—102 (1908). — Die Entwicklungsgeschichte der Keimdrüsen von Rana esculenta. Ein Beitrag zum Sexualitätsproblem. Festschrift zum 60. Geburtstag R. Hertwigs, Bd. 2, S. 61—224. 1910. — **Kynoch, J. A. C.:** Repeated ovariotomy. J. Obstetr. **2**, 366—371 (1902). — **Levi, G.:** Sull'origine della cellule sessuali. Monit. zool. ital. **15**, 244—246 (1904). — Sull'origine delle cellule germinali. Arch. di Fisiol. **2**, 243—245 (1905). — Lesioni sperimentali sull'abbozzo urogenitale di larve die Anfibi e loro effetti sull'origine della cellule sessuali. Arch. Entw.mechan. **19**, 295—317 (1905). — **McCosh, G. K.:** Origin of germ cells in Amblystoma maculatum. Anat. Rec. **41**, 78 (1928) (Abstract). — The origin of the fat-bodies in Amblystoma maculatum. Anat. Rec. **45**, 109—119 (1930a). — The origin of the germ cells in Amblystoma maculatum. J. Morph. a. Physiol. **50**, 569—612 (1930b). — **McGregor, J. H.:** The spermatogenesis of Amphiuma. J. of Morph. **15**, Suppl. 57—104 (1899). — **Marshall, A. M.:** On certain abnormal conditions of the reproductive organs in the frog. J. Anat. a. Physiol. **18**, 121—144 (1884). — **Meredith, W. A.:** Pregnancy after removal of both ovaries for dermoid tumor. Brit. med. J. **1**, 1360 (1904). — **Mihálkovics, G. von:** Untersuchungen über die Entwicklung des Harn- und Geschlechtsapparates der Amnioten. Internat. Mschr. Anat. u. Histol. **2**, 41—64, 65—106, 284—306, 307—385, 387—433, 435—486 (1885). — **Morgan, C. L.:** Animal life and intelligence. London: Edward Arnold 1891. — **Morgan, A. H.,** and **MacNab, A.:** The effects of ovariectomy in Diemyctylus viridescens. Anat. Rec. **34**, 128 (1926) (Abstract). — **Morris, M. M.:** Pregnancy following removal of both ovaries and tubes. Boston med. J. **144**, 86—87 (1901). — **Nussbaum, M.:** Zur Differenzierung des Geschlechts im Tierreich. Arch. mikrosk. Anat. **18**, 1—121 (1880). — Zur Entwicklung des Geschlechts beim Huhn. Verh. anat. Ges. 15. Verslg Bonn 1901, 38—40. — Über den Einfluß der Jahreszeit, des Alters und der Ernährung auf die Form der Hoden und Hodenzellen der Batrachier. Arch. mikrosk. Anat. u. Entw.gesch. **68**, 1—121 (1906). — **Obreshkove, V.:** Accessory testicular lobes in

Diemyctylus viridescens, their probable origin and significance. J. Morph. a. Physiol. **39**, 1—46 (1924). — **Okkelberg, P.**: The early history of the germ cells in the brook lamprey, Entosphenus wilderi (Gage), up to and including the period of sex differentiation. J. of Morph. **35**, 1—152 (1921). — **Owen, R.**: On parthenogenesis. London 1849. — **Pallot, G.**: A propos de la regénération ovarienne et des modifications périodique de l'epithelium vaginal chez le Rat blanc. C. r. Soc. Biol. Paris **99**, 1333—1334 (1928). — **Papanicolaou, G. N.**: Ovogenesis during sexual maturity as elucidated by experimental methods. Proc. Soc. exper. Biol. a. Med. **21**, 393—396 (1924). — **Parkes, A. S.** and **Bellerby, C. W.**: Studies on the internal secretions of the ovary. I. The distribution in the ovary of the oestrus-producing hormone. J. of Physiol. **61**, 562—575 (1926). — **Parkes, A. S., Fielding, U.** and **Brambell, F. W. R.**: Ovarian regeneration in the mouse following complete double ovariotomy. Proc. roy. Soc. B **101**, 328—354 (1927). — **Pencharz, R. I.**: Experiments concerning ovarian regeneration in the white rat and white mouse. Wistar Inst. Bibl. Service, **1929a**, Nr 124. — Experiments concerning ovarian regeneration in the white rat and white mouse. J. of exper. Zool. **54**, 319—342 (1929b). — **Perle, S.**: Origine de la première ébauche génitale chez Bufo vulgaris. C. r. Acad. Sci. Paris **184**, 303—304 (1927). — **Reagan, F. P.**: Some results and possibilities of early embryonic castration. Anat. Rec. **11**, 251—267 (1916). — **Reinhard, L.**: Die Entwicklung des Parablast und seine Bedeutung bei Teleostiern nebst der Frage über Entstehung der Urgeschlechtszellen. Arch. mikrosk. Anat. u. Entw.-mechan. **103**, 339—356 (1924). — **Richards, A., Hulpieu, H. R.** and **Goldsmith, J. B.**: The restudy of the germ-cell history in the fowl. Anat. Rec. **34**, 158 (1926) (Abstract). — **Richards, A.** and **Thompson, J. T.**: The migration of the primary sexcells of Fundulus heteroclitus. Biol. Bull. Mar. Labor. biol. Wood's Hole **40**, 325—348 (1921). — **Robinson, A.**: The formation, rupture, and closure of ovarian follicles in ferrets and ferret-polecat hybrids, and some associated phenomena. Trans. roy. Soc. Edinburgh **52**, 303—362 (1918). — **Rubaschkin, W.**: Zur Frage von der Entstehung der Keimzellen bei Säugetierembryonen. Anat. Anz. **32**, 222—224 (1908a). — Über das erste Auftreten und Migration der Keimzellen bei Vögelembryonen. Anat. H. **35**, 241—261 (1908b). — Über die Urgeschlechtszellen bei Säugetieren. Anat. H. **39**, 603—652 (1909). — Über das erste Auftreten und Migration der Keimzellen bei Säugetierembryonen. Anat. H. **41**, 243 (1910a). — Chondriosomen und Differenzierungsprozesse bei Säugetierembryonen. Anat. H. **41**, 399—431 (1910b). — Zur Lehre von der Keimbahn bei Säugetieren. Über die Entwicklung der Keimdrüsen. Anat. H. **46**, 343—411 (1912). — **Rückert, J.**: Über die Entstehung der Excretionsorgane bei Selachiern. Arch. Anat. u. Entw.gesch. 205—278, 1888. — **Ruthven, A. G., Thompson, C.** and **Gaige, H. T.**: The herpetology of Michigan. Mich. Handbook. Ser. 1928, Nr 3 (Univ. Mus., Univ. of Mich.). — **Schapitz, R.**: Die Urgeschlechtszellen von Amblystoma. Ein Beitrag zur Kenntnis der Keimbahn der Urodelen Amphibien. Arch. mikrosk. Anat. **79**, 41—78 (1912). — **Semon, R.**: Studien über den Bauplan des Urogenitalsystems der Wirbeltiere. Dargelegt an der Entwicklung dieses Organsystems bei Ichthyophis glutinosus. Jena. Z. Med. u. Naturwiss. **26**, 89—203 (1891). — **Sharp, L. W.**: An introduction to cytology. McGraw-Hill Book Company, New York 1926. — **Simkins, C. S.**: On the origin and migration of the so-called primordial germ cells in the mouse and the rat. Acta zool. (Stockh.) **4**, 241—284 (1923). — Origin of the germ cells in Ecteinascidia. J. Morph. a. Physiol. **39**, 295—321 (1924). — Origin of the germ cells in Trionyx. Amer. J. Anat. **36**, 185—214 (1925). — Origin of the sex cells in man. Amer. J. Anat. **41**, 249—294 (1928). — **Sink, E. W.**: The origin of the germ cells in the toadfish (Opsanus tau). Mich. Acad. Sci. Rep. **14**, 212—215 (1912). — **Spehl, G.** et **Polus, J.**: Les premiers stades du développement des glands génitales chez l'Axolotl. Archives de Biol. **27**, 63—90 (1912). — **Stieve, H.**: Die Entwicklung der Keimzellen und der Zwischenzellen in der Hodenanlage des Menschen. Z. mikrosk.

anat. Forsch. **10**, 225—285 (1927). — **Stromsten, F. A.:** History of the germ cells in the goldfish. Anat. Rec. **44**, 254 (1929) (Abstract). — **Sun, Y. C.:** Post-pubertal ovogenesis in the guinea-pig. Anat. Rec. **25**, 114—115 (1923) (Abstract). — **Swezy, O.:** The ovarian chromosome cycle in a mixed rat strain. J. Morph. a. Physiol. **48**, 445—474 (1929). — **Swezy, O.** and **Evans, H. M.:** Maturation of human embryonic ova. Proc. Soc. exper. Biol. a. Med. **27**, 10 (1929). — Ovogenesis in the mammals. Proc. Soc. exper. Biol. a. Med. **27**, 11 (1929b). — **Swift, C. H.:** Origin and early history of the primordial germ cells in the chick. Amer. J. Anat. **15**, 483—516 (1914). — Origin of the definitive sex-cells in the female chick and their relation to the primordial germ cells. Amer. J. Anat. **18**, 441—470 (1915). — Origin of the sex-cords and definitive spermatogonia in the male chick. Amer. J. Anat. **20**, 375—410 (1916). — **Swingle, W. W.:** The germ cells of anurans. I. The male sexual cycle of Rana catesbeiana larvae. J. of exper. Zool. **32**, 235—331 (1921). — Is there a transformation of sex in frogs? Amer. Naturalist **56**, 193—210 (1922). — Germ cell and germ gland development in male Rana catesbeiana tadpoles. Anat. Rec. **24**, 381—382 (1923) (Abstract). — Sex differentiation in the bullfrog (Rana catesbeiana). Amer. Naturalist **59**, 154—176 (1925). — The germ cells of anurans. II. An embryological study of sex differentiation in Rana catesbeiana. J. Morph. a. Physiol. **41**, 441—546 (1926). — **Tschaschin, S.:** Über die Chondriosomen der Urgeschlechtszellen bei Vögelembryonen. Anat. Anz. **37**, 597—607, 621—631 (1910). — **Valaoritis, E.:** Über die Oogenesis beim Landsalamander (Salamandra maculata). Zool. Anz. **2**, 597—599 (1879). — **Vanneman, A. S.:** The early history of the germ cells in the armadillo, Tatusia novemcincta. Amer. J. Anat. **22**, 341—364 (1917). — **Waldeyer, W.:** Eierstock und Ei. Ein Beitrag zur Anatomie und Entwicklungsgeschichte der Sexualorgane. Leipzig: Wilh. Engelmann 1870. — Die Geschlechtszellen. Handbuch der vergleichenden und experimentellen Entwicklungslehre der Wirbeltiere von O. Hertwig, Bd. 1, S. 86—474. 1901—03. — **Weismann, A.:** Die Kontinuität des Keimplasmas als Grundlage einer Theorie der Vererbung. Jena 1885. — **Wheeler, W. M.:** The development of the urinogenital organs of the lamprey. Zool. Jb. **13**, 1—88 (1899). — **Willier, B. H.:** The development of implanted chick embryos following the removal of the „primordial germ cells." Anat. Rec. **34**, 158 (1926) (Abstract). — **Wilson, E. B.:** The cell in development and heredity. New York: Macmillan Company 1925. — **Wilson, K. M.:** Origin and development of the rete ovarii and the rete testis in the human embryo. Contrib. Embryol., Carnegie Inst. Washington Publ. **17**, 69—89 (1926). — **Winiwarter, H. von:** Contribution à l'étude de l'ovaire humain (I. Appareil nerveux et phéochrome. II. Tissu musculaire. III. Cordons médullaires et corticaux). Archives de Biol. **25**, 683—757 (1910). — **Winiwarter, H. von** et **Sainmont, G.:** Über die ausschließlich postfetale Bildung der definitiven Eier bei der Katze. Anat. Anz. **32**, 613—616 (1908a). Nouvelles recherches sur l'ovogenèse et l'organogenèse des mammifères (chat). Archives de Biol. **24**, 1—142, 165—276, 373—431, 627—650 (1908b). — **Witschi, E.:** Experimentelle Untersuchungen über die Entwicklungsgeschichte der Keimdrüsen von Rana temporaria. Arch. mikrosk. Anat. **85**, 9—113 (1914). — Studies on sex differentiation and sex determination in amphibians. I. Development and sexual differentiation of the gonads of Rana sylvatica. J. of exper. Zool. **52**, 235—266 (1929a). — Studies on sex differentiation and sex determination in amphibians. II. Sex reversal in female tadpoles of Rana sylvatica following the application of high temperature. J. of exper. Zool. **52**, 267—292 (1929b). — **Woodger, J. H.:** Observations on the origin of the germ-cells of the fowl (Gallus domesticus), studied by means of their Golgi bodies. Quart. J. microsc. Sci. **69**, 445—462 (1925). — **Woods, F. A.:** Origin and migration of the germ-cells in Acanthias. Amer. J. Anat. **1**, 307—320 (1902). — **Wyhe, J. W. van:** Über die Mesodermsegmente des Rumpfes, und die Entwicklung des Excretionssystems bei Selachiern. Arch. mikrosk. Anat. **33**, 461—516 (1889).

	Seite
v. **Cholnoky, B.**, Beiträge zur Kenntnis der Karyologie von Microspora stagnorum. Mit 26 Textabbildungen	707
Pattri, Hermann Otto Erich, Über die Doppelbrechung der Spermien. Mit 12 Textabbildungen	723
Haas, G., Über drüsenähnliche Gebilde der Epidermis am Kopfe von Typhlops braminus. Mit 5 Textabbildungen	745
Lange, Heinz-Herbert, Die Phagocytose bei Chironomiden. Mit 21 Textabbildungen	753
Autorenverzeichnis	806

Ergebnisse der Biologie.

Herausgegeben von **K. v. Frisch**-München, **R. Goldschmidt**-Berlin-Dahlem, **W. Ruhland**-Leipzig, **H. Winterstein**-Breslau. Redigiert von **H. Winterstein**-Breslau.

Achter Band. Mit 88 Abbildungen. V, 372 Seiten. 1932.
RM 36.—; gebunden RM 38.60

Der Winterschlaf. Von Dr. D. Ferdmann und Dr. O. Feinschmidt-Charkow.

Das Ciliensystem in seiner Bedeutung für Lokomotion, Koordination und Formbildung mit besonderer Berücksichtigung der Ciliaten. Von Dr. Bruno M. Klein-Wien.

Nestbau und Brutpflege bei Amphibien. Von Professor Dr. W. Wunder-Breslau.

Grundlinien einer allgemeinen Ökologie der Diatomeen. Von Dr. R. W. Kolbe-Berlin-Dahlem.

Namen- und Sachverzeichnis. — Inhalt der Bände I—VIII.

Siebenter Band. Mit 109 Abbildungen. X, 724 Seiten. 1931.
RM 77.—; gebunden RM 79.80

The Permeability of the Erythrocyte. Von Professor Dr. M. H. Jacobs-Philadelphia Pa. (U. S. A.).

Vergleichende Betrachtung der pathologischen Anatomie und Physiologie des Zentralnervensystems. Von Privatdozent Dr. Ludwig Singer-München.

Brutpflege und Nestbau bei Fischen. Von Professor Dr. W. Wunder-Breslau.

Das Determinationsproblem. Von Professor Dr. O. Mangold-Berlin-Dahlem. Dritter Teil: **Das Wirbeltierauge in der Entwicklung und Regeneration.**

Die chemischen Vorgänge beim biologischen Kohlehydratabbau. Von Privatdozent Dr. Karl Wetzel-Leipzig. Erster Teil: **Die einleitenden Prozesse der biologischen Zuckerspaltung.**

Elektive Vitalfärbungen. Probleme, Ziele, Ergebnisse, aktuelle Fragen und Bemerkungen zu den Methoden. Von Privatdozent Dr. Josef Gicklhorn-Prag.

Namen- und Sachverzeichnis. — Inhalt der Bände I—VII.

VERLAG VON JULIUS SPRINGER IN BERLIN

VERSTÄNDLICHE WISSENSCHAFT

Vom Zellverband zum Individuum. Von Dr. Otto Steche,
Professor der Zoologie, Leipzig. Mit 72 Abbildungen. VIII, 160 Seiten. 1929. Band X. Gebunden RM 4.80 (abzüglich 10% Notnachlaß)

Das Buch von Steche wirft die Frage nach dem Wesen und der Bedeutung des eigenen „Ich" auf. Es behandelt, von dem Leben und den Funktionen der einzelnen Zelle ausgehend, die Entwicklung der ersten primitiven Zellverbände mit dem Ziel, als letzten großen Zellstaat den Menschen in diese Reihe einzufügen. Das Buch bespricht alle Fragen in rein naturwissenschaftlicher Art, ist aber so leicht und einfach geschrieben und bietet so viele Illustrationen, daß es für jeden Leser verständlich und interessant sein wird.

Aus den Werkstätten der Lebensforschung. Von Dr.
Paul Weiss, Wien. Mit 11 Abbildungen. V, 192 Seiten. 1931. Band XII. Gebunden RM 4.80

Der Band zeigt, wie der moderne Forscher arbeitet, wie in seinen Werkstätten gearbeitet wird. Er berichtet über seine Methoden bis zum letzten, bis zur Mitteilung in der Form einer wissenschaftlichen Arbeit. Dargestellt ist dieser Forschungsweg an Hand von Beispielen aus dem Gebiet der Biologie.

Schlaf und Traum. Von Professor Dr. Hans Winterstein, Direktor
des Physiologischen Instituts der Universität Breslau. Mit 22 Abbildungen. V, 135 Seiten. 1932. Band XVIII. Gebunden RM 4.80

Die Rätsel von Schlaf und Traum haben schon seit den ältesten Zeiten die Phantasie der Menschen beschäftigt. Für ein Drittel unseres Lebens tauchen wir in die geheimnisvollen Tiefen des Schlafes, aus denen wir nur kärgliche Trümmer von Traumerinnerungen an das Licht des Wachens retten. Was geht in diesem seltsamen Zustand vor? Was ist sein Wesen, sein Sinn und sein Ursprung? Wir sind auch heute noch weit entfernt, den dunklen Vorhang heben zu können, aber die wissenschaftliche Forschung hat doch von verschiedenen Standpunkten aus da und dort einen Blick hinter die Kulissen zu werfen vermocht. Die Ergebnisse dieser Forschung leicht faßlich zu schildern ist die Absicht des Verfassers in dem neuen Bändchen.

Die Welt der Sinne. Eine gemeinverständliche Einführung
in die Sinnesphysiologie. Von **W. v. Buddenbrock,** Professor der Zoologie an der Universität Kiel. Mit 55 Abbildungen. VI, 182 Seiten. 1932. Band XIX. Gebunden RM 4.80

Die Aufgabe der „Verständlichen Wissenschaft" ist es, ihre Leser, von den verschiedensten Gesichtspunkten ausgehend, mit der sie umgebenden Welt und ihren Gesetzen bekannt zu machen. Da ist es natürlich von besonderem Reiz, wenn wir uns einmal intensiver mit den naturgegebenen Werkzeugen beschäftigen, mit denen allein wir Naturbeobachtung treiben können. Ehe wir uns mit dem Mikroskop und ähnlichen technischen Hilfsmitteln eingehend befassen, sollten wir alle versuchen, uns über die Fähigkeiten und Leistungen unserer Sinne ein klares Bild zu machen, die ja nicht nur für das „Was", sondern auch für das „Wie" der Auffassung unserer Umwelt entscheidend sind.

VERLAG VON JULIUS SPRINGER IN BERLIN

MIX
Papier aus verantwortungsvollen Quellen
Paper from responsible sources
FSC® C105338

If you have any concerns about our products,
you can contact us on
ProductSafety@springernature.com

In case Publisher is established outside the EU,
the EU authorized representative is:
**Springer Nature Customer Service Center GmbH
Europaplatz 3, 69115 Heidelberg, Germany**

Printed by Libri Plureos GmbH
in Hamburg, Germany